# Philosophical Reflections on Neuroscience and Education

Bloomsbury Philosophy of Education

Edited by Michael Hand
*Bloomsbury Philosophy of Education* is an international research series
dedicated to the examination of conceptual and normative questions
raised by the practice of education.

Forthcoming in the series:

*Cherishing and the Good Life of Learning*, Ruth Cigman

Also available in the series:

*A Critique of Pure Teaching Methods and the Case of Synthetic Phonics*,
Andrew Davis

Also available from Bloomsbury:

*Authority and the Teacher*, William H. Kitchen
*Early Childhood and Neuroscience*, Mine Conkbayir

# Philosophical Reflections on Neuroscience and Education

William H. Kitchen

Bloomsbury Academic
An imprint of Bloomsbury Publishing Plc

B L O O M S B U R Y
LONDON • OXFORD • NEW YORK • NEW DELHI • SYDNEY

**Bloomsbury Academic**
**An imprint of Bloomsbury Publishing Plc**

| | |
|---|---|
| 50 Bedford Square | 1385 Broadway |
| London | New York |
| WC1B 3DP | NY 10018 |
| UK | USA |

**www.bloomsbury.com**

**BLOOMSBURY and the Diana logo are trademarks of Bloomsbury Publishing Plc**

First published 2017

© William H. Kitchen, 2017

**British Library Cataloguing-in-Publication Data**
A catalogue record for this book is available from the British Library.

ISBN: HB: 978-1-4742-8369-4
ePDF: 978-1-4742-8371-7
ePub: 978-1-4742-8372-4

**Library of Congress Cataloging-in-Publication Data**
A catalog record for this book is available from the Library of Congress.

Cover design by Clare Turner

Series: Bloomsbury Philosophy of Education

Typeset by Deanta Global Publishing Services, Chennai, India
Printed and bound in Great Britain

To find out more about our authors and books visit www.bloomsbury.com. Here you will find extracts, author interviews, details of forthcoming events and the option to sign up for our newsletters.

'... if the conceptual framework is awry, then incoherence ensues. Questions that make no sense will be asked. Experiments will be designed to answer questions that make no sense. The results of experiments will be misunderstood and misinterpreted.'

P.M.S. Hacker

# Contents

# Series Editor's Foreword

Bloomsbury Philosophy of Education is an international research series dedicated to the examination of conceptual and normative questions raised by the practice of education.

Philosophy of education is a branch of philosophy rooted in and attentive to the practical business of educating people. Those working in the field are often based in departments of education rather than departments of philosophy; many have experience of teaching in primary or secondary schools; and all seek to contribute in some way to the improvement of educational interactions, institutions or ideals. Like philosophers of other stripes, philosophers of education are prone to speculative flight, and the altitudes they reach are occasionally dizzying; but their inquiries begin and end on the ground of educational practice, with matters of immediate concern to teachers, parents, administrators and policymakers.

Two kinds of question are central to the discipline. *Conceptual* questions have to do with the language we use to formulate educational aims and describe educational processes. At least some of the problems we encounter in our efforts to educate arise from conceptual confusion or corruption – from what Wittgenstein called 'the bewitchment of our intelligence by means of language'. Disciplined attention is needed to such specifically educational concepts as learning and teaching, schooling and socializing, training and indoctrinating, but also to the wider conceptual terrain in which educational discourse sits: what is it to be a person, or to have a mind, or to know or think or flourish, or to be rational, intelligent, autonomous or virtuous? *Normative* questions have to do with the justification of educational norms, aims and policies. What educators do is guided and constrained by principles, goals, imperatives and protocols that may or may not be ethically defensible or appropriate to the task in hand. Philosophers of education interrogate the normative infrastructure of educational practice, with a view to exposing its deficiencies and infirmities and drawing up blueprints for its repair or reconstruction. Frequently, of course, the two kinds of question overlap: inappropriate aims sometimes rest on conceptual muddles and our understanding of educational concepts is liable to distortion by ill-founded pedagogical norms.

In terms of scholarly output, philosophy of education is in rude health. The field supports half a dozen major international journals, numerous learnt societies and a busy annual calendar of national and international conferences. At present, however, too little of this scholarly output finds a wider audience, and too few of the important ideas introduced in journal articles are expanded into fully developed theories. The aim of this book series is to identify the best new work in the field and encourage its authors to develop, defend and work out the implications of their ideas, in a way that is accessible to a broad readership.

It is hoped that volumes in the series will be of interest not only to scholars and students of philosophy of education and neighbouring branches of philosophy, but also to the wider community of educational researchers, practitioners and policymakers. All volumes are written for an international audience: while some authors begin with the way an educational problem has been framed in a particular national context, it is the problem itself, not the local framing of it, on which the ensuing arguments bear.

Michael Hand,
University of Birmingham, UK

# Preface

The education–neuroscience experiment is one massive error! That is the overriding sentiment of what you are about to read. I make no apologies for this bleak and overtly negative outlook. Indeed, there is little positive to say about the neuroscience venture into educational discourse at all; except maybe to show us how collaborative research ought not to be conducted.

This is not a new movement. It has been building towards its peak for quite some time (since the mid-1990s at least) and we arrive now at a stage when the collaboration is rife within many universities across the world.

Intuition tells us that these disciplines – neuroscience and education – should surely have an interest in one another. Both, it seems, have a vested interest in the brain and how it works. It seems at least convenient, therefore, for neuroscience to share its findings with education, since, after all, knowing more about the brain may well give new insights into how best to learn.

However, intuition, as is often the case, leads intellect a merry dance. Our seemingly intuitive considerations seem to supersede our intellectual faculties, and we begin to see how this major collaborative effort is founded on some weak conceptual grounds.

I recall attending a teachers' conference in my native Northern Ireland, at which the presenter – a hard-line proponent of brain-based learning – led the audience on an 'intuitive' walk through the workings of the brain. He offered the following opening to his talk: 'Please finish these sentences … (1) The ear … (2) The stomach … (3) The eye … (4) And finally, the brain … .'

You will be unsurprised, I imagine, to learn that the answers which returned in unison from the audience were thus: (1) 'hears'; (2) 'digests' (a few murmurs of 'eats' could also be heard!); (3) 'sees', and finally, (4) various answers like 'learns' or 'thinks'.

Now, only a philosopher – which I reluctantly admit to being – would find fault with three of these statements. Does the ear hear? Does the eye see? And, most notably for our discussions in this book, does the brain learn/think?

I suggest that most readers at this point will answer 'yes' to all three questions. By the end of this book, I consider it my duty to address these conceptual problems, to delineate between sense and nonsense, and to show the reader that these questions – let alone the seemingly intuitive answers to them – are so profoundly nonsensical that the answers which follow are nothing more than mindless drivel. In other words, I hope to convince you that we cannot utter such phrases as 'The ear hears', 'The eye sees' or 'The brain learns' without being taken for a fool. Quite the task, but one which I trust you will agree, is worth reading on for!

Once we wrestle with these intuitions, we begin to see the appeal of neuroscience unfold in the educational setting. Once we unravel the conceptual confusions which underpin the fusion of these two disciplines, we see there is little residue worth saving.

By the end of the book, I hope you will agree. But in the absence of agreement, at least I hope to offer some interesting considerations and reflections which need resolving before moving any further with this failed educational neuroscience experiment.

Dr W. H. Kitchen
30 November 2016

# Acknowledgements

There are, as expected, a number of people who I must thank for the production of this book.

First, to my intellectual mentor and PhD supervisor Dr Hugh Morrison, who guided me on this venture during my studies at University, and indeed entrusted me with this project after having committed a reasonable amount of his own working life to it. These ideas are not only mine; they were developed within him, and manifest through me. I owe him a great deal.

To my friend and colleague Dr Noel Purdy, another one of my PhD supervisors, whose wisdom and insight will always be greatly appreciated. He is an unusual beast to find in academia, given his kindness and courtesy, which is often missing within our work. He gives willingly of his time and expertise, and has afforded me a great deal more interest than many others may have done. For that, and so much else, I am in his debt.

Finally, as always to my wife and family, who always encourage me in these obscure and wacky ventures! They are subjected, more than anyone else, to my nuances and foibles, and I have never felt anything other than their encouragement and support.

# Wittgensteinian Abbreviations

It is common practice when quoting Wittgenstein's philosophical works to cite his texts using standard abbreviations, followed by a proposition number rather than a page number, denoted using the symbol '§', with the year of publication omitted. For a detailed overview of all of Wittgenstein's writings, and an insight into the citation style, the reader is referred to Glock (1996: 3–8).

The following is the list of abbreviations of Wittgenstein's texts cited throughout:

*TLP*   Wittgenstein, L. (1922). *Tractatus Logico-Philosophicus.* D. F. Pears and B. F. McGuinness (Trans). Oxon: Routledge & Kegan Paul.

*RFM*   Wittgenstein, L. (1956). *Remarks on the Foundations of Mathematics.* G. E .M. Anscombe and G. H. von Wright (Eds), G. E. M. Anscombe (Trans). Oxford: Blackwell.

*Z*   Wittgenstein, L. (1967). *Zettel.* G. E. M. Anscombe and G. H. von Wright (Eds), G. E. M. Anscombe (Trans). Oxford: Blackwell.

*OC*   Wittgenstein, L. (1969). *On Certainty.* G. E. M. Anscombe and G. H. von Wright (Eds), Denis Paul and G. E. M. Anscombe (Trans). Oxford: Blackwell.

*CE*   Wittgenstein, L. (1976). *Cause and Effect: Intuitive Awareness.* R. Rhees (Ed), P. Winch (Tran), *Philosophia,* 6, 392–445.

*RPP I*   Wittgenstein, L. (1980a). *Remarks on the Philosophy of Psychology: Volume I.* G. E. M. Anscombe and G. H. von Wright (Eds), G. E. M. Anscombe (Tran). Oxford: Blackwell.

*RPP II*   Wittgenstein, L. (1980b). *Remarks on the Philosophy of Psychology: Volume II.* G. E. M. Anscombe and G. H. von Wright (Eds), G. E. M. Anscombe (Tran). Oxford: Blackwell.

*CV*   Wittgenstein, L. (1980c). *Culture and Value.* G. H. von Wright and H. Nyman (Eds), P. Winch (Tran). Oxford: Blackwell.

*LW I*   Wittgenstein, L. (1982a). *Last Writings on the Philosophy of Psychology: Volume I.* G. H. von Wright and Heikki Nyman (Eds), C. G. Luckhardt and Maximilian A. E. Aue (Trans). Oxford: Blackwell.

*LW II*      Wittgenstein, L. (1992). *Last Writings on the Philosophy of Psychology: Volume II.* G. H. von Wright and Heikki Nyman (Eds), C. G. Luckhardt and Maximilian A. E. Aue (Trans). Oxford: Blackwell.

*WLC I*      Wittgenstein, L. (1982b). *Wittgenstein's Lectures: Cambridge 1930-1932.* D. Lee (Ed.). Chicago: Chicago University Press.

*WLC II*     Wittgenstein, L. (1982c). *Wittgenstein's Lectures: Cambridge 1932-1935.* A. Ambrose (Ed). Chicago: Chicago University Press.

*PI*      Wittgenstein, L. (2009). *Philosophical Investigations (4th Revised Edition).* P. M. S. Hacker and Joachim Schulte (Eds), G. E. M. Anscombe, P. M. S. Hacker and Joachim Schulte (Trans). Oxford: Wiley-Blackwell.

# Introduction

Neuroscience and education seem to be natural partners. Neuroscience is the growing science of the day, and education has a vested interest in what neuroscience investigates; namely, the brain. Surely if we understand the brain more, we will understand the 'stuff' of education more? This seems to make much sense.

How learning happens is the business of any educator, and making learning more likely to happen is the focus of all educational endeavours. The axiology of education is largely irrelevant in the pursuit of this one unifying goal – learning. Schools are supposed to cultivate learning, teachers are supposed to make it happen and students are supposed to be educated not only on subject content, but also on how they learn. Learning *how* to learn is the modern focus of education, and students must acquire this lifelong skill if they are to hope to be successful at learning. It is insufficient to simply teach students maths, reading and how to use a computer; they must be taught *how to learn*.

This shift within education has led to an interesting collision between neuroscience and education, since the minutiae of both disciplines seems to contain ever-increasing overlaps. But we must pause for a moment's reflection to ensure that this pursuit is well founded and that our intuitions have not run amuck. We must ensure that the practical research is established within sound conceptual analysis and afforded a robust philosophical critique. That will be the focus of this book. You will find no interviews, nor any real case studies in any meaningful educational sense. This is, perhaps, a tad uncommon for a book which will examine, in the first instance, an overtly practical science like neuroscience, and an even more practical social science like education.

The philosophical method which is invoked in this book is so-called *a priori*, ordinary language philosophy, the focus of which, as Hacker (2014: 5) notes, is centred around establishing a coherent conceptual framework within which one can conduct one's *a posteriori* investigations: 'If the conceptual framework is awry, then incoherence ensures. Questions that make no sense will be asked.

Experiments will be designed to answer questions that make no sense. The results of experiments will be misunderstood and misinterpreted.'

I will cover four broad themes in this book. First, it is important to set the argument in the context of where the neuroscience–education collaboration began, and what the rationale for its inception has been. This will be entirely descriptive.

Secondly, there is a need to critique this approach to education and subject both neuroscience in general, and its educational derivations in particular, to a robust conceptual and philosophical examination. I will introduce three core arguments into this debate at this stage: namely, the mereological fallacy (Bennett and Hacker 2003), the first-person/third-person category error and the notion of irreducible uncertainty.

Thirdly, I will examine the overlapping principles between neuroscience, Cartesian philosophy and mind-brain identity theory, to show that neuroscience is nothing more than a modernized materialist version of ill-fated and problematic seventeenth-century mentalist philosophy, serving only to replace the concept of 'mind' with the concept of 'brain'. Extensive efforts will be made at this stage to show that the concepts of neuroscience are not nearly as anti-Cartesian as they ought to be, and I will further the case of an adoption of Wittgenstein's solutions to these problems, in his consideration of the so-called inner/outer picture.

Finally, I will show in the final major theme of the book that intrinsic models of education (of which neuroscience and its educational derivations are examples) are untenable, and that these models must be replaced by relational models, which are founded on philosophical concepts borrowed from quantum physics. As a result of this philosophical shift, I will argue for the replacement of Cartesian–Newtonian ideas within educational philosophy, with Wittgensteinian–Bohrian ideas. This new model of education will then be applied to resolve some educational paradoxes which have plagued education for many years.

Part One

# An Introduction to Neuroscience and Education

1

# Neuroscience, Brain-based Learning and Education

## 1.1 A definition of neuroscience

It is particularly difficult to capture a satisfactory definition of a discipline as diverse as neuroscience. Nevertheless, a broad definition must be established for the bounds of this book, in particular with reference to education.

Bennett and Hacker (2003: 2) suggest that 'cognitive neuroscience operates across the boundary between two fields, neurophysiology and psychology'. In a sense, cognitive neuroscience serves as the experimental science that makes empirical efforts to link brain properties with behaviour. The remit of the discipline seems to be to explain behaviour by attaining a fuller understanding of how the brain works.

More specifically, cognitive neuroscience will be examined in the bounds of this book, in particular to examine the role which it can (or cannot) play with reference to education. In one of the first major works of the cognitive neuroscience movement – *Cognitive Neuroscience: The Biology of the Mind* – one of the fathers of the movement, Michael Gazzaniga and two other cognitive neuroscientists, Richard Ivry and George Mangun, outline in the preface of the text that 'cognitive neuroscience is taking the scientific community by storm' (Gazzaniga, Ivry and Mangun 1998: xiii), and offer the following insight as to what cognitive neuroscience offers to science:

Scientists now realize that studying the mind's complex processes – perception, language, attention, memory, control of movement, feelings, and consciousness itself – has become a task that is not only scientifically tractable, but is approachable by cognitive and neural means. The disciplines of cognitive psychology, behavioural neurology, and neuroscience now feed off each other, contributing a new view to the understanding of the mechanisms of the mind. This development has led to the emergence of the field of cognitive neuroscience.

Various aspects of the definition are important to note. First, there is a (perhaps troubling) interchangeability between the concept of 'mind' and the concept of 'brain'. Indeed the authors of this book seemed content to use the terms in place of one another freely, seemingly identifying one with the other without concern.

Secondly, these authors seem to have attributed a range of abilities to the mind, such as 'perception, language, attention, memory, control of movement, feelings, and consciousness itself' and outline that an examination of the *brain* and its neurophysiology will lead to a more detailed understanding of how the *mind's* processes work. This was, in truth, the basis for the birth of neuroscience in these early stages, something which requires closer examination.

For the moment, let it be clear that Gazzaniga, Ivry and Mangun (1998) have given significant reason to question the credentials of neuroscience as credible science. Gazzaniga, Ivry and Mangun (1998: xiii) set out to show that 'the brain enables the mind'. This was what these authors saw, at their time of writing, to be the aim of cognitive neuroscience. Much of that aim remains intact to this day. This position is curious and requires critique.

Consider and compare the more recent publication *Principles of Cognitive Neuroscience*, written in 2013, which offers the following basic definition of cognitive neuroscience:

> Cognitive neuroscience is a relatively new discipline that has arisen from the recent marriage of neuroscience, a biomedical field that has flourished both conceptually and technically during the past century, and cognitive science, a field of study rooted in the long-standing interest of natural philosophers and psychologists in understanding human mental processes. Consistent with these progenitors, research on cognitive neuroscience integrates investigations of brain structure and function, and seeks to measure cognitive abilities and behavior to understand how the human brain works at all levels. (Purves et al. 2013: 1)

At first glance, it appears that some of the potential difficulties with the definition Gazzaniga, Ivry and Mangun (1998) offered have been sufficiently well mastered in this more up-to-date book. Indeed, even on the concept of the mind, the authors note that the '*mind* is a notoriously difficult term to define' (Purves et al. 2013: 2). These authors continue:

> Cognitive neuroscience is defined by the work at the intersection of cognitive science and neuroscience. Thus, cognitive neuroscientists must have grounding in both these domains. They must be able to think about the cognitive processes that shape our behavior and the contents of our mental lives, and understand cognitive psychology and related fields. But they must also be able to relate those

processes and theories to underlying brain function, which requires proficiency in the key findings and the tools of neuroscience. Cognitive neuroscience thus combines all the difficulties of measuring brain function with all the problems of trying to accurately assess cognition and behaviour, as well as the complexities of trying to link them together. (2013: 9)

Taking these two definitions together – Gazzaniga, Ivry and Mangun (1998) and Purves et al. (2013) – it should be eminently clear that the intersection between cognitive processes and neurophysiological processes is the interest of the cognitive neuroscientist. In light of this definition, it also becomes clearer what the appeal of cognitive neuroscience might be for education: namely, to find a foothold in a deeper and more scientific understanding of the underlying precepts of the cognitive processes of interest to education, such as learning, thinking and understanding. This author asks: 'Can cognitive neuroscience deliver on these promises in an educational context?' This question will be examined throughout this book.

## 1.2  The neuroscientific method

What, then, of the cognitive neuroscientific method? If connections and relations are to be established between the brain and the mind, and the mind's processes are to be better understood, how are the cognitive neuroscientists going to go about it?

The experimental methodology of neuroscience has developed over the short lifespan of the discipline; however the staple of the neuroscientific method is the neural scan, which comes in various forms: functional magnetic resonance imaging (fMRI), magnetic resonance imaging (MRI), positron emission tomography (PET), electroencephalography (EEG) and event-related potentials. The neural scan, in whatever form, should highlight how the neurophysiology of the brain 'enables' the mind to carry out its processes. Even those neuroscientists who are not necessarily mind-brain identity theorists would contest that an examination of the brain will lead to a better understanding of what the *brain* controls. So, even if the brain does not *enable* the mind as Gazzaniga, Ivry and Mangun (1998) posit, it is still seen as the centre or the hub of perception, memory, feelings, consciousness and the like. This view will be contested later in this book. In particular, in recent years the view that the brain is the learning, thinking organ, or that learning and thinking take place inside the brain, has taken hold within educational discourse, at least in part due to the recent involvement and contributions of neuroscience in educational thinking.

The neuroscientific method, therefore, encompasses a collection of neural imaging techniques, which are, it seems, used to delve 'inside' the skull to attain insights into the neurophysiology of the brain. The view from within neuroscientific circles which this book seeks to examine is whether these neurophysiological 'states' can be *conceptually* linked to psychological states, with particular reference to the psychological attributes of interest to education, such as learning, thinking and understanding. As Bennett and Hacker (2003: 2) observe, 'The logical or conceptual relations between the physiological and the psychological are problematic.' This is the motivation for the critique of neuroscience in this book, particularly in relation to education.

## 1.3   A definition of neuroscience from within education

Neuroscience is a broad discipline, ever widening in recent years due to the increased interest and alleged importance of understanding the workings of the brain in the various activities which are of interest to most human beings. Geake and Cooper (2003) suggest that there are around 50,000 neuroscience researchers in Europe alone, with a similarly impressive number in the United States (100,000), displaying the significant research interests in the functioning of the brain and the implications that such functioning might have on human behaviours and activities.

Interestingly, Geake and Cooper (2003: 8–9) suggest the following definition of neuroscience, set in the context of what should be of interest to educational discourse:

> Cognitive neuroscience is a wide field embracing a rich variety of experimental paradigms and approaches, from the bimolecular to the behavioural. … Areas of experimental interest include vision, spatial cognition, audition and music, emotions, memory, motor function, language, and consciousness, most (if not all) of which can inform our understanding of cognitive behaviours relevant to education, for example, intelligence, learning, memory, motivation, literacy, creativity.

It seems, therefore, that Geake and Cooper attribute a major significance to neuroscience in a broad number of areas, and for the purposes of this discussion, they feel that neuroscience has a significant role to play in a more explicit understanding of the cornerstones of education, which they have listed as 'intelligence, learning, memory, motivation, literacy, creativity'. The appeal of neuroscience for education, therefore, appears to be that it purports to be capable of providing answers to some of education's most vexing questions surrounding learning, memory and intelligence. Later in this chapter, the inherent value of neuroscience to education will be examined, with particular reference made to how realistic and viable any collaboration between neuroscience and education might be.

## 1.4  A definition of Brain-based learning

Brain-based learning is a collection of learning 'theories' which purport to explain learning through developing a fuller understanding of the role the brain plays in learning. It is a study which calls on evidence from educationalists, cognitive psychologists and neuroscientists to support its claims. According to Caine and Caine (1991) and Gulpinar (2005: 302–3), Brain-based learning theory encompasses twelve main principles, listed as follows:

1. All learning engages the entire physiology;
2. The brain/mind is social;
3. The search for meaning is innate;
4. The search for meaning occurs through patterning;
5. Emotions are critical to patterning;
6. The brain/mind processes parts and whole simultaneously;
7. Learning involves both attention and peripheral perception;
8. Learning is both conscious and unconscious;
9. There are at least two approaches to memory (rote learning system, spatial/contextual/dynamic memory system);
10. Learning is developmental;
11. Complex learning is enhanced by challenge and inhibited by threat associated with helplessness and fatigue;
12. Each brain is uniquely organized.

This definition provides a standard understanding of brain-based learning models; although alternative brain-based learning theories might differ slightly from others. Nevertheless, the attraction for appealing to brain-based learning models is apparent: there is a systematic attempt within brain-based learning theories to explain how learning occurs, and how best to teach on the basis of this explanation; claims which are substantiated in evidence from mind/brain philosophy, Gestalt psychology, cognitive psychology and neuroscience.

Jensen (2008: xii), suggests that keeping pace with the fast-moving nature of neuroscience and brain research in general is one of the challenges facing modern-day educationalists. Moreover, the same author suggests that modern-day education is founded on a new knowledge of the brain and its workings, which previous generations of educationalists were not so fortunate to have. This, he suggests, is why embracing the brain-based learning movement is so important:

> Based on research from the disciplines of neuroscience, biology, and psychology, our understanding of the relationship between learning and the brain now encompasses the role of emotions, patterns, meaningfulness, environments, body rhythms, attitudes, stress, trauma, assessment, music, movement, gender, and enrichment. By integrating what we now know about the brain with standard educational practices, *Brain-Based Learning* suggests that schools can be transformed into complete learning organizations. (Jensen 2008: xii)

Jensen (2008: xiii) attempts to bolster this case, claiming that 'we are all great natural learners', and continues to opine that brain-based learning may serve as the saviour to education:

> When students are provided with a learning environment that is optimal for learning, graduation rates increase, learning difficulties and discipline problems decrease, a love for learning flourishes, administrators focus on the real issues, and learning organizations thrive. In short, creating an organization around the way the *brain naturally learns* [this author's italics] best may be the simplest and most critical educational reform ever initiated. In fact, of all the reforms, nothing provides a better return on your investment of time, energy, and money than developing a brain-based approach to learning.

In summary, if brain-based learning is adopted, attainment improves, problems disappear and a 'love' for learning flourishes. The 'learning organization' is the aim, and the student will thrive in this environment which is founded on an appreciation of how the 'brain learns'. Money is saved, and our investment is safe. These are bold claims. We would be correct to demand sound evidence and rigorous methods for their foundation. This book seeks to address these issues.

## 1.5  Some internal warnings: Dispelling Neuromyths

Some neuroscientists with an interest in education (Goswami 2006; Tallis 2010; Howard-Jones 2010; Rose 2013) are keen to dispel so-called neuromyths from infesting in education and other disciplines. Some of these authors may warn that some of Jensen's (2008) claims are neglectful of more careful approaches.

Indeed, in the above excerpt, this author has emphasized one point of particular interest: namely, the line 'brain naturally learns'. Rather unfortunately it is common practice for the brain-based learning movement to posit that the brain is the learning organ, and that learning environments, organizations and curricula should be shaped around this learning brain. Indeed, as Jensen (2008:2) notes, with reference to the brain, 'the vast complexity of our "thinking organ" has left scholars short of an efficient explanation of how it works'. Such approaches have given birth to notions like Brain Gym, Fingerprint Learning, Learning Styles and other fashionable educational 'tricks' which attempt to found themselves within neuroscience to give themselves scientific credibility. Some more careful neuroscientists have therefore responded and warned that such ploys are, in fact, not at all founded in neuroscientific research.

For the time being, however, let it be clear what brain-based learning has as its core aim. To quote Jensen (2008: 4) one last time: 'Brain-based education considers how the brain learns best. The brain does not learn on demand by a school's rigid, inflexible schedule. It has its own rhythms.' This tenet of brain-based learning will be shown to be the work of science fiction rather than science fact. Neuroscience, it seems, is often used to provide a cover of credibility to brain-based learning, giving the impression that the concept of the learning, thinking brain is a modern-day scientific marvel. A quick search on the Amazon.co.uk website (which hardly constitutes research, but it does give a feel for how increasingly popular this field of brain-based learning is) will return over 1,100 books on brain-based learning. These ideas have permeated the mainstream thinking within education. They are rife within educational discourse throughout the world. And they are all predicated on the belief that the best way to educate children is to consider how their 'brains learn best'.

## 1.6 The rationale for Brain-based learning and neuroeducation

Educational models and theories have historically been placed into one of two camps. First, there is support for explanations of learning to be found *within* the locality of the learner, first-person monistic theories of learning, generally encompassed (deliberately or otherwise) under the broad philosophical doctrine of Cartesianism and further espoused by Lockean philosophy.

Cartesianism, broadly speaking, suggests the existence of an inner mechanism (perhaps housed inside the mind/brain), which guides learning in a local, causal model, that is, all the necessary information within the locality of the learner. Learning is to be understood by understanding the inner mechanism which guides.

The theories based on Cartesian philosophy take an atomistic view of the learner. Moreover, such theories have been variously advanced in an educational sense most recently in the development of brain-based learning theories and neuroeducational investigations. Indeed, such approaches tend to search for explanations within the brain, generally holding the view that a fuller understanding of the brain and its workings will lead to subsequent fuller understandings of the way people learn, understand and operate in an educational setting.

Alternatively, theories of learning which purport that learning is best understood by rejecting the notion of an 'inner guidance', in favour of behavioural analysis *only*; so-called third-person monistic theories of learning encompassed under the educational doctrine of behaviourism. Behaviourism suggests that all psychological phenomena are reducible to behavioural phenomena, and that abilities such as learning are to be understood *only* through behavioural analysis. Such theories were historically espoused within education in the late nineteenth and early twentieth centuries by Pavlov, Watson, Skinner and company, to further the notion that behaviour was all that really mattered in understanding how the human being functions.

The educational trend is to jump between these two far-reaching educational doctrines and their respective derivations.

These educational obsessions have manifested themselves as a series of attempts to establish what learning is *precisely*, and how it is best to achieve successful learning outcomes within the educational setting. The answer to this most fundamental of educational questions has led many researchers to what they believed to be the answer that would stand the test of time.

The most modern version of this search for answers within educational discourse has led to the development of so-called 'brain-based learning' theories and neuroeducation. The search for the justification and fullest understanding of how and why learning takes place, and how precisely teaching and learning are to be most successful has led to the development of the notion that learning is to be housed inside the brain or that the activity of learning is to be ascribed to the brain.

Brain-based learning theories have thus given rise to the development of applying neuroscience to education. The concatenation of the disciplines of neuroscience and education has developed into studies such as 'neuroeducation' and 'mind, brain and education'; that is, the disciplines which call on neuroscience, at least in part, for their scientific underpinnings in a quest to answer educational questions[1].

The general claims of brain-based learning and neuroeducational studies, which are of interest for this discussion, tend to fall into one of two (or perhaps both) camps:

1. First, there is the claim that the brain is the *agent* of psychological predicates, insomuch as it is claimed that it is *the brain* that learns, thinks, understands, sees and so on.
2. This differs slightly from the second claim of brain-based learning and neuroscientific (and subsequently neuroeducational) theories, which purport that the brain is the *locus* of psychological predicates, suggesting that learning, thinking, understanding, seeing, pattern-searching and so on, all take place *inside* the brain.

The distinction between these two claims will be made clearer later in Part 2. In any case, however, the claim of brain-based learning theories and neuroeducational studies is, essentially, that psychological predicates can be ascribed to the brain either as an agent or as a locus (or both). Therefore, in order to critique brain-based learning and neuroscientific approaches to education, it suffices to critique their central claims. It is the purpose of Part 2 to do precisely that. Before this can be done, however, it seems sensible to clarify further the educational rationale for brain-based learning and neuroscientific approaches, citing examples and showing what precisely the remit of neuroscience is in the educational context. Once the rationale has been established, the aim is to show that such attempts to solve educational quandaries through neuroscience are fraught with inherent difficulties.

## 1.7  Geake's five arguments for a collaboration between neuroscience and education

In an early development in the neuroeducation discussion, neuroeducationalist John Geake offers five arguments which he believes are reason to give serious

---

[1]  The terms 'neuroeducation' and 'brain-based learning' are, it seems, fundamentally connected; neuroeducation is, more precisely speaking, the brain-based learning approach founded in neuroscientific principles. That is, it is the educational doctrine which purports to explain learning and inform teaching practices and pedagogy through neuroscientific principles.

consideration to a potentially fruitful collaboration between neuroscience and education. He argues that we should search for a 'mutual middle way' (Geake 2005: 10) in an attempt to coalesce the ideas of neuroscience and education into a working model for educational practice, arguing that 'the education profession could benefit from embracing rather than ignoring cognitive neuroscience. Moreover, educationalists should be actively contributing to the agenda of future brain research. That is, a cognitive neuroscience-education nexus should be a two-way street' (Geake 2005: 12).

His justification for such a potential collaboration begins with citing a much-cited claim by John T. Bruer:

> We send our children to school to learn things they might not learn without formal instruction so that they can function more intelligently outside school. If so, recommendations for school reform should explicitly appeal to and implement our best, current understanding of what learning and intelligence are. In the public debate on school reform, this is seldom the case. Common recommendations – raising standards, increasing accountability, testing more, creating markets in educational services – are psycho atheoretical, based at best on common sense and at worst on naïve or dated conceptions of learning. (Bruer 1994: 273)

Geake is prudent to cite Bruer in this instance, insomuch as Bruer himself is correct to claim that children are sent to school to attain a type of learning that they would otherwise be deprived of if no such formal schooling were offered.

Furthermore, both Bruer and Geake sensibly claim that education in general, and learning in particular, should be guided, wherever possible, by the most up-to-date version of coherent scientific practice. There remains a serious question, however, as to whether or not neuroscience is a stable science on which to predicate educational reform. Indeed, Geake himself warns that his ideas were not 'without caveats' (Geake 2005: 10), encouraging readers to pursue his ideas with a degree of caution and scepticism.

Geake further offers five main arguments to support the thesis that there is at least rationale for pursuing a collaborative body of work between neuroscience and education.

1. *The in-principle argument* – the notion that since humans are biological entities and since behaviour is also biological in nature it follows that the 'stuff of education is neurobiological' (Geake 2005: 10).
2. *The professional imperative argument* – the idea that since there is 'considerable media interest in educational applications of brain functioning

research' and that 'school teachers are interested in brain functioning relevant to learning and development in children' (Geake 2005: 10–11), it follows that there is a professional duty on educationalists to take an interest in neuroscience and what it has to offer to the debate, to ensure a scientifically well-informed approach is taken to education (Geake 2005: 11)

3. *The in-practice argument* – the notion that since existing knowledge of the workings of the brain have already made contributions to how we operate within education, that such understanding is likely only to be increased with future collaborative work.

4. *The self-interest argument* – the idea that educationalists *ought* to listen, by virtue of the fact that they should have a self-interest in improving their practice, and neuroscience is well positioned to offer scientific support to any likely improvements.

5. *The opportunistic argument* – finally, the idea that since so much research time, effort and finance have been invested in neuroscience in the current research world, education would be foolish as a discipline not to capitalize on this wave of interest. The fact, therefore, that neuroscience has established rising credentials within research communities, leads Geake (and others[2]) to argue that education as a profession 'could benefit from embracing rather than ignoring cognitive neuroscience' (Geake 2005: 12).

Whether we agree or disagree with the basis of Geake's (2005) five arguments, there is no doubt that these views underpin a rationale for pursuing the collaboration, and indeed form the basis for widely held views from within both disciplines for pursuing the collaboration. However, the credibility of the collaboration requires further examination, and the seemingly intuitive underpinning links between these disciplines require a stern review.

---

[2] It should be noted, however, that authors such as Steven Rose and Raymond Tallis, both practising neuroscientists, warn that such a view is predicated on so-called brain-hysteria, and that it ought to, in fact, be avoided.

# Collaborative Reports in Neuroscience and Education

## 2.1 Major collaborations between neuroscience and education

Although it is widely acknowledged that education and neuroscience have been in a 'closet' collaboration for around the last two decades (Howard-Jones et al. 2007:3),[1] only recently has the collaboration become more prominent, largely due to large-scale research taking place to endorse such a collaboration as potentially fruitful for both disciplines. This author has already demonstrated in the previous section that the modern-day interest in such collaboration has been championed by authors from 2003 onwards, with more vigour and interest than was perhaps publically displayed in the first instance.

More recently – in fact, as recently as January 2014 and February 2015 – major inter-disciplinary and collaborative reports are being published as to the importance of neuroscience in educational discussions and considerations about teaching and learning, and the role which the brain plays in teachers' interactions with their pupils (Howard-Jones 2014; Howse 2015). Further developments are transpiring on a regular basis, with claims of new neuroscientific insights into how memories are shaped and formed, how so-called episodic memory works and how such memories are stored, and how nerves firing and synapses fusing contribute to these process (Mundasad 2015; Webb 2015).

Furthermore, there are increasingly more online resources[2] around the general area of *mind, brain and education*, with claims on the credibility of neuroscientific findings for educational purposes, which are purported to be of practical use to

---

[1]  Henceforth referred to as Teaching and Learning Research Programme (TLRP) (2007)
[2]  See: Fingerprint Learning (NI), http://www.fingerprintlearning.com/; Learning Zone (UK), http://learning.imascientist.org.uk/; Dana Foundation (International), http://www.dana.org/

teachers in their daily lives in assisting the teacher to work with the 'grain of the brain' and to find 'unique learning' solutions for all of their learners.

There is an enthusiastic purpose about much of the modern neuroscientific and educational inter-disciplinary research, with much hope harboured of the collaboration being useful both in the short and the long terms. As it has already been highlighted, heightened interest in brain research, and so-called brain-hysteria has swept the international research community, leading to many prominent neuroscientists and educationalists putting their case forward for brain-based, neuro-centred approaches to learning in particular and education in general.

This section will be dedicated to showing precisely how this case has been made, by giving a detailed examination of an example of such an argument for the neuroscience–education collaboration. The example in question will be a detailed exegesis of the Royal Society's *Brain Waves Module 2: implications for education and lifelong learning*.

As a pre-emptive note, however, the author refers the reader to other major collaborative works in neuroscience and education, most notable being the aforementioned TLRP (2007), led by prominent neuroeducationalist Paul Howard-Jones, in which early foundations are set for further investigation; foundations which are clearly built upon in the Royal Society's (2011) work. As such, the Royal Society document is taken as a reliable paradigmatic example of what a collaborative report looks like in the surge towards forging a fruitful neuroscience–education collaboration.

## 2.2 The Royal Society's endorsement and recommendations for neuroscience and education

On the backdrop of the arguments put forward in Geake and Cooper (2003), Geake (2005), TLRP (2007) and other related documents, the brain-based learning movement (particularly the study of neuroeducation) has become significantly more prominent in recent years since the Royal Society decided in 2011 to endorse the application of neuroscience for educational purposes when it commissioned the project entitled, 'Brain Waves Module 2: implications for education and lifelong learning.'

This project was aimed at broadening the scope of neuroscientific studies, particularly in module 2, to encompass, as part of its remit, an informed glance at how these neuroscientific principles could impact educational thinking. Its findings had built on the groundwork laid down by authors like those previously cited.

Various leading authorities in the neuroscientific and educational worlds put their names to the project to give it the credibility it required to permeate the mainstream thinking within educational discussions. Indeed, the fact that it was sanctioned and subsequently endorsed by the Royal Society ensured that the project was to be taken seriously as a document which could, eventually, serve as the beginnings of a radical educational reform in terms of educational thinking, practices and policy.

Despite a pre-emptive warning of caution from within their own ranks about the potential scope of their project (The Royal Society 2011: v), there are many bold assertions about what is achievable in the neuroscience–education collaboration if it is pursued with full verve. There is a suggestion of understanding learning and developmental disorders such as Attention Deficit Hyperactivity Disorder (ADHD) and dyslexia, on the basis of grasping the associated neuroscience of such disorders. Further, there are hopeful claims of establishing the neuroscientific principles which make learning possible, and there is talk of how to teach more successfully to harness these learning dispositions; claims which are based on the notion that the brain is the learning organ. There is talk of 'synapses fusing' and 'neurons firing' and 'connections being made', to give an altogether scientific feel to the learning process. Learning, in this way, is seen as a process which finds its natural home inside the brain; a notion with undoubted intuitive appeal.

There are also many cases of educational romanticism in suggesting that neuroscience can offer insight into how the *brain* can learn; information which would be utterly inaccessible in the absence of the neuroscientist who can shed light on such issues. Every teacher is to be aware of these neuroscientific principles, if they are to be considered as serious in their quest to provide a top-class education. Neuroscience, therefore, is seen as the necessary tonic to save education from itself. By equipping education with a neuroscientific wing, learning can be understood via an understanding of the brain and its workings, and teaching can be tailored to ensure learning is given the chance to be more successful. At least, that is how the story goes. Indeed,

> Neuroscience is the empirical study of the brain and the connected nervous system. The brain is the organ that enables us to adapt to our environment – in essence, to learn[3]. Neuroscience is shedding light on the influence of our genetic make-up on learning over our life span, in addition to environmental factors. This enables us to identify key indicators for educational outcomes, and provides a scientific basis for evaluating different teaching approaches. (The Royal Society 2011: 3)

---

[3]   This seems like a rather restrictive definition of learning, but we shall proceed reluctantly.

Such claims are a development of previous works, like Goswami (2006: 3) who argues that 'our understanding of how to optimize the brain's ability to benefit from teaching' is always growing on the basis of neuroscientific findings.

## 2.3 Curbing the enthusiasm

The remainder of this chapter will be dedicated to showing that the Royal Society in particular, and the neuroeducation community in general have failed in establishing the boundaries of their work. They have overestimated their own value, and have awarded themselves a self-attributed ability to impact on education in ways that their science and their conceptual framework in no way support. Indeed, they estimate that their influence over education can be comparable (eventually) to the impact which medical sciences made towards medical practices over a century ago (The Royal Society 2011: v); a startling suggestion, made without warrant or foundation.

The claim is that since the brain is the 'learning organ', to understand it is to understand how people learn and subsequently how best to teach. The claim that the brain is either the agent or the locus of learning (or any psychological predicate or attribute for that matter, such as thinking, understanding and so on) is, despite its intuitive appeal, open to criticism, as I will show. Indeed, talk of brain's learning, or of learning taking place inside the brain will be dismissed later in the book as incoherent and conceptually flawed.

Whether there is scope for neuroscience to inform education *at all* is of secondary importance to whether or not the questions that are being posed within neuroscience, and subsequently apply to neuroeducation and brain-based learning, for example, make sense, and whether or not the conceptual foundations of the science are sound. In the absence of such sound foundations, the collaboration is doomed to fail before it even begins.

## 2.4 Four major recommendations

For the moment, however, it seems sensible to further develop an understanding of what, precisely, the Royal Society proposes neuroscience can do for education. In the 'Brain Waves Module 2' project, there are four major recommendations made with regard to the impact that neuroscience can have on education and teacher training (giving rise to the alleged importance of neuroeducation as a collaborative discipline). They are:

1. Neuroscience should be used as a tool in educational policy;
2. Training and Continued Professional Development (CPD) should include a neuroscientific element in relation to relevant educational issues, particularly Special Educational Needs (SEN);
3. Neuroscience should inform adaptive learning technology; and
4. Knowledge exchange should be increased.

(The Royal Society 2011: 19–21)

Let it be clear in the process of critiquing these recommendations that this author does not seek to undermine the elements of the Royal Society's report which have addressed many of the intricate problems which have plagued neuroscientific discourse for many years. Indeed, other publications – such as Geake (2005), TLRP (2007) and Howard-Jones (2010) – set about making clear where neuroscience has gone wrong in the past, arguing that the neuromyths which have hindered neuroscience should be discarded in order for the discipline to be taken seriously.

This, however, is where this author parts with the Royal Society (2011), and other such publications. Indeed, it is puzzling why such efforts would be expended to recalibrate the boundaries of sense within neuroscience, only to make bold and unfounded assertions about its applicability to other disciplines, such as education. If neuroscience is in a time of cautious optimism (Geake 2005: 10; Goswami 2006; The Royal Society 2011: v), this should be reflected in its adoption to education; unfortunately, this is not the case.

Indeed, given that the authors of the Royal Society (2011: v, emphasis added) report acknowledge that neuroscience presents 'opportunities as well as *challenges* for education' and that the discipline of neuroscience is still relatively infantile in nature, insomuch as the authors 'urge caution in the rush to apply so-called brain-based methods, many of which do not yet have a sound basis in science', it seems strange that this document ends with a recommendation that neuroscience should be used to inform educational policy, and that teachers should be trained in neuroscientific principles.

The move by the Royal Society to endorse this project demonstrates the widespread belief within prominent research communities that the collaboration between neuroscience and education is one which is likely to work. This author, however, remains sceptical of such claims, and makes the following four counter-recommendations:

## 2.5  Four counter-recommendations

1. *In response to recommendation (1): 'Neuroscience should be used as a tool in educational policy';*
This recommendation is, as Bruer (1997) warned many years ago, at the beginning of the neuroscience–education collaboration, 'a bridge too far'. Furthermore, given that Geake and Cooper (2003), Geake (2005), Goswami (2006), TLRP (2007) and the Royal Society (2011) themselves have suggested that caution is needed in the collaboration between neuroscience and education, it seems strange that the Royal Society seems to relent on this warning of caution to make bold and strong assertions like this one.

The justification for this recommendation is, in fact, predicated on nothing other than the view that there is an opportunity for neuroscience to inform education, and now seems the right time to grasp this opportunity. Neuroscience tells us about the brain, and education, it is contested, is interested in the workings of the brain, so it seems, at least on face value, that this collaboration is destined to work.

In fact, the credibility of the rationale which the Royal Society adopts brings us no further than previous arguments made by, for example, Geake (2005), who, as it has been shown, offered five weak arguments for the pragmatic adoption of neuroscientific principles for educational purposes.

This author offers a simple counter-recommendation: only if – and that's a big 'if' – neuroscience can be shown to be scientifically credible, conceptually coherent and – perhaps most importantly – amenable to collaborative work with education, can it be offered as a tool for educational policy. In the absence of such credentials, it should be rejected. This author will spend much of Part 2 of this book showing that the conceptual foundations of neuroscience in general, and neuroeducation in particular are questionable at best. As such, any suggestion that neuroscience should be used as a tool in educational policy looks altogether irresponsible.

2. *In response to recommendation (2): 'Training and Continued Professional Development (CPD) should include a neuroscientific element in relation to relevant educational issues, particularly Special Educational Needs (SEN)';*
The notion that teachers should be trained in neuroscientific principles is no more credible than the previous recommendation. Indeed, even if the conceptual credibility of neuroscience could be established, the value of the study to education is far from clear.

This distinction is captured clearly in the Scottish philosopher David Hume's famous 'is-ought' distinction for scientific discussions. Essentially Hume (1739) argued that there is a fundamental flaw in how so-called 'is' statements often lead to mistaken 'ought' statements. What this means is, there is a failure to recognize the difference between saying how things 'are' and how things 'ought' to be on the basis of how things are.

The nature of 'is' statements is that they are, by definition, empirical. 'Ought' statements, on the other hand, are normative in nature. The connection between them, Hume argues, is often overlooked, and as a result one finds that there are a great many 'ought' statements offered on the back of 'is' statements, with no apparent connective or further justification offered.

If Hume is correct, how things 'are' in neuroscience need not say anything about how things 'ought' to be in education. In other words, the normative values of education need not be governed by the empirical findings of neuroscience, regardless of the apparent common interests shared by the two disciplines.

The further notion that neuroscience should be used 'particularly for SEN' is more than a little strange. Indeed, it is commonplace to hear of neuroscience, or perhaps more generally the brain sciences, offering so much by the way of insight into matters of SEN, in particular in areas such as developmental disorders, for example autism and ADHD. The reality is, though, if neuroscience is unfit for general educational use, then it is also unfit for matters concerning SEN. So, the same arguments apply here as before; namely, the conceptual framework adopted by neuroscience must be sound, and its questions must be cogent, otherwise the only natural consequence is a transgression on the bounds of sense. In other words, the empirical findings of neuroscience are of secondary importance to the conceptual grounds on which such questions are built.

The consequence of this is another simple counter-recommendation: it would be entirely irresponsible to train teachers in neuroscientific principles, given that the application of neuroscience to education has not yet been established. In fact, it remains entirely unclear whether neuroscience has anything of worth to offer to education in general, as this author will show in Part 2, and as such it would be unjustified to encourage teachers to think about the neuroscientific approaches to education. Regarding the use of neuroscience for informing SEN matters in education, this author contests that such a move is equally unjustified and warns that the use of the SEN example as a case for the adoption of neuroscience is, in fact, careless.

3. *In response to recommendation (3): 'Neuroscience should inform adaptive learning technology';*

It is believed that 'new educational technologies provide opportunities for personalized learning that our education system cannot otherwise afford' (The Royal Society 2011: 20).

The notion of adaptive learning is, essentially, the use of technological devices and computers to assist with learning, in such a way that the learning is tailored, or 'adapted' to the learner's needs on the basis of the learner's inputs into the device.

The advantage of such uses of technology in education, it is believed, can 'open up learning opportunities outside the classroom and hence improve access to those currently excluded from education in adulthood and later life' (The Royal Society 2011: 20).

The role that neuroscience can play in providing these educational opportunities is largely a collaborative one, between the neuroscientific community, the educational community and technological industries (ibid). The insights from neuroscience can, it is argued, be used to 'inform the design of educational technologies' (ibid).

This author contests, on the contrary that any use of adaptive learning technologies is fraught with inherent complications, which are outlined clearly in Morrison (2014). The essence of Morrison's (2014) rejection of the use of adaptive learning technologies in general is predicated on the notion that such models are built inside a Newtonian paradigm of education, a notion which this author will return to in Parts 2 and 4 of this book.

Morrison's (2014) general claim, however, is that 'adaptive learning [has] no scientific merit'. Citing Suter (1989), Oppenheimer (1955) and Stapp (1993), Morrison (2014) concludes, 'Opponents of [adaptive learning models] seem to be driven by a deep-rooted sense that there is something unsettling about this project. Surely learning is a human activity and "data science" cannot replace the teacher and the school?'

Furthermore, Morrison (2014) also argues that adaptive learning models are committed to a conceptual model (Newtonianism) and a methodology (so-called Item Response Theory) which are unsuitable for education. Therefore, despite the fact that they may have some use as an additional resource for teachers, their pupils, and parents in the home environment, their scientific credibility, and thus their widespread use within education as a pedagogical tool, is extremely questionable.

This author, therefore, counter-recommends that any notion of a wholesale adoption of adaptive learning technologies, whether informed by neuroscience or not, should be avoided. Invoking Morrison (2014), this author contends that adaptive learning models are unsuitable for educational use, demonstrating further that something is awry within the conceptual foundations of the neuroscience–education collaboration if such pedagogical tools are being encouraged from within this collaboration.

4. *In response to recommendation (4): 'Knowledge exchange should be increased'.* This author recommends that knowledge exchange is possible between disciplines only when the disciplines are conceptually amenable to one another. This is far from a matter of fact with regard to neuroscience and education. Indeed, this author poses the question: Are the *concepts* of neuroscience and the *concepts* of education open to fusion? Subsequently, one must ask: Are neuroscience and education compatible? These questions are pertinent, and there are many complexities which remained unanswered.

More specifically, it is important that the manner in which the concepts of purported common interest to neuroscience and education are equivalent across the disciplines before any knowledge exchange takes place. That is to say, if neuroscientists mean something different by 'learning', for example, than their educational counterparts, then knowledge exchange between the disciplines is ill-advised. Much work remains to be done, before neuroscience and education could even be viewed as compatible, let alone beginning to dictate each other's research agendas.

# A Local Paradigmatic Example, Founded on an International Research Phenomenon

## 3.1 Setting the scene

To set the Royal Society's claims in some sort of paradigmatic context, it seems sensible to give an illustration – by way of a practical example – of how some of the neuroscientific principles, which were endorsed by the Royal Society, have already permeated educational reform. Such a paradigmatic example of a curriculum founded in neuroscientific principles can be found in the guise of the Northern Ireland Revised Curriculum.

It is interesting to note that, even though the Royal Society, by their own admission suggested that further work was required before neuroscientific principles were to be adopted as the educational norm (in 2011), the Council for the Curriculum, Examinations and Assessment (CCEA) had already adopted much of what the Royal Society proposed, in their own *Pathways* documents, written in 2003. Once again, the fact that such radical educational reform was proposed on the basis of neuroscientific insight – a discipline which is still regarded (even by its leading proponents) to be in its infancy – displays a lack of seriousness given to the claim that educationalists should remain cautious about the wholesale application of neuroscience to education.

## 3.2 Neuroscience and curriculum reform

CCEA appealed to neuroscience to add scientific support to their adopted learning theory of constructivism, which was to be applied in Northern Ireland's schools. CCEA cite new knowledge of how the 'brain works' among their reasons for changing the curriculum and taking different approaches to assessment and pedagogy. In the section entitled 'Rationale for the revised curriculum and assessment proposals' of the *Pathways* documents, the following appears:

> Recently neuroscience has established a number of factors, which are critical to learning and to motivation, about how our brains process information. We now know that the human brain creates meaning through perceiving patterns and making connections and that thought is filtered through the emotional part of the brain first. The likelihood of understanding taking place is therefore increased significantly if the experience has some kind of emotional meaning, since the emotional engagement of the brain on some level is critical to its seeing patterns and making connections. ... Neuroscience, therefore, highlights the need for learning to be emotionally engaging, particularly during the 11–14 range when so much is going on with adolescents to distract them from school. (CCEA 2003b: 22)

These claims encapsulate all that *seems* to be attractive about neuroscience and its application to education. The theme that appears to run throughout all of the claims is that neuroscience gives rise to a fuller understanding of the brain, consequently giving a fuller understanding of how learning can be successful. Inherent in this claim is that if educators understand the brain, they get an understanding of learning for free. There is talk of an understanding of how 'brains process information', and how they 'make connections'. The claim, therefore, must insinuate that if educators can understand *how* the brain 'processes information' and 'makes connections' then teaching methodologies can be tailored to ensure learning is more likely to be successful.

Moreover, there is a major claim regarding the requirement of emotional engagement within learning, since, it is argued, engagement of the 'emotional brain' is 'on some level critical' to successful learning. That is, learners should be emotionally engaged through the teaching to which they are exposed, in order to enhance the chances of subsequent learning being successful. Similar difficulties emanate from this argument, as outlined earlier in the chapter. For example, what precisely does this neuroscientific insight tell us about how teachers are supposed to teach? Are all lessons to be an emotional performance, stirring the creative juices of every learner to ensure that there is an emotional connection between the learner and what is to be learnt? In fact, what is meant by 'emotionally engaging', in relation to teaching is left altogether ambiguous. Moreover, what impact is such a claim to have on the nature and structure of *what* is taught. So, for example, how is a mathematics teacher to make algebra 'emotionally engaging'? Perhaps it is a sign of the educational times that the emotional engagement of a subject is given more consideration than the *content* of the subject. More worryingly, pseudoscientific evidence is now being used to support such a paradigm shift within education.

Furthermore, talk of these neuroscientific facts being 'critical to learning', even when leading proponents of neuroscience are adamant that the neuroscientific discipline has yet to be firmly established, gives an overwhelming feeling that these ideas *must* be adopted, lest education lose out.

The argument also, outlined above, that 'so much is going on with adolescents to distract them from school', bears all the hallmarks of scare-mongering. At a time when education is calling out for a saviour, many are appealing to neuroscience as a solution. There is no doubt that adolescents are easily distracted, but it is wishful thinking, at best, that neuroscientific approaches to education can resolve this issue, and there exists *no* evidence to support this claim in any way.

### 3.2.1 The structure of the curriculum: Under attack

Neuroscience is also used to make the case for educational reform regarding how the curriculum is to be structured. Developing the idea that learning is to be emotionally engaging to the learner, CCEA also cite neuroscientific evidence which suggests that the compartmentalization of the curriculum into discrete subjects should be replaced by a more thematic framework:

> Our current emphasis on learning within separate subject disciplines dates back at least a century and is based on the notion that each subject is a distinct form of knowledge with separate characteristics, concepts and procedures which encourage efficient learning. Over the last decade, we have begun to learn more about how the brain processes information. … We are beginning to question the wisdom of compartmentalising learning whilst expecting young people to cope with multi-dimensional problems. (CCEA 2003c: 2)

Within this claim, there are various issues which give rise to serious questions about the rationale upon which neuroeducational reform is predicated. First, there is an acknowledgement that subjects are currently compartmentalized because each has a 'distinct form of knowledge'. The fact, therefore, that a change is proposed, must surely suggest that neuroscientific insight can offer some evidence to reject this notion of subject 'distinctiveness'? That is, there must be some form of neuroscientific evidence which refutes the notion that school subjects are distinct forms of knowledge. The only argument offered in this instance is the claim that young people are required now to cope with 'multi-dimensional problems'. Precisely what is meant by 'multi-dimensional problems' in an educational context is not entirely clear.[1]

Moreover, even if this premise is accepted, it remains unjustified to suggest that the best way to approach such problems is to ensure that the curriculum is thematic as opposed to compartmentalized. The suggestion is that, since neuroscience has found that the brain 'processes information' in such a way that it 'seeks patterns' and 'searches for connections' in its efforts to learn, the emphasis on education is subsequently to make learning opportunities more 'connected'; in essence, to clear the way for the brain to make the 'connections' which it is searching for. In this way, the proposed 'structure' and 'workings' of the brain, borrowed from neuroscience, are used educationally to justify a change in curriculum structure.

### 3.2.2 'Authentic' learning

Finally, CCEA claim that since the brain searches for patterns and connections in the way that has been proved neuroscientifically, it automatically follows that learning opportunities should be provided using collaborative project work to ensure that work is contextualized and emotionally engaging – what CCEA call 'authentic learning'. By calling such learning 'authentic' CCEA seek to undermine alternative forms of learning, established in traditional teaching methods, which they purport to be outdated and dogmatic in nature. Neuroscience therefore is used to add scientific clout to the claim that learning should be made 'authentic' through participatory, collaborative group-work situations:

> Recent brain research indicates that the brain searches for patterns and inter-connections as its way of making meaning. Researchers theorise that the human brain is constantly searching for meaning and seeking patterns and connections. Authentic learning situations increase the brain's ability to make connections and retain new information. When we set the curriculum in the context of human experience, it begins to assume a new relevance. (CCEA 2003c: 3)

These claims bear all the hallmarks of attempts at fitting an educational model around pseudoscientific claims and hypotheses. Talk of 'authentic learning' is far-fetched and unfounded. There is no offer of evidence to support the claim

---

[1] Perhaps 'multi-dimensional' refers to the nature of so-called 'real-life' problems, which make reference to some kind of real-life experience. Such problems may include more than one 'dimension', that is, may call on solutions from a broad range of subjects or topics.

that these so-called 'authentic learning' situations can play a role in increasing successful learning outcomes. Rather, these claims and subsequent half-attempts at establishing the claims in scientific evidence only serve to display that the reform process began with an attractive outcome, and the 'evidence' was doctored to fit the mould.

What might the implications be for the next generation of learners, if all their learning situations are required to be 'authentic' before they are permitted as acceptable? Are we to dispense with algebra because it cannot be made 'real' to the learners? When a student asks his teacher in class 'why are we doing this?' and the teacher cannot rationalize their teaching with CCEA's 'authenticity', is the teaching to be abandoned as useless? Or, more sensibly, should formal schooling be viewed as the exposure to types of learning that would otherwise be unattainable to the student if there were no such formal schooling on offer? That is, to place a restriction on formal education that all learning is to be 'authentic' and 'real' for learners, on the basis of pseudoscientific claims, is to deprive learners of the vast array of *other* educational and learning opportunities which can be accessed in the absence of such a restriction.

Moreover, the claim that authenticity in learning is a prerequisite for more successful learning outcomes is unfounded. To claim that authentic learning opportunities increase the brain's ability to retain information is to suggest that the brain has the ability to distinguish between 'useful' and 'useless' information at the point of learning. Claims of 'filters' in the brain which have the ability to distinguish between 'useful' and 'useless' information, or claims that the brain can identify an 'authentic' learning opportunity from an 'unauthentic' one, would be radical if they were capable of being proved or any evidence could be provided to support them.

The brain, however, is not that kind of entity; and so to predicate educational reform on the basis of the brain being something that it is not is grossly irresponsible. The brain is not an 'authenticity filter'. If it were, some of the greatest ever abstract mathematical, literature, historic and scientific breakthroughs of all time would have been unattainable. Indeed, in what way was Fermat's Last Theorem 'authentic', when he postulated its mathematical validity in 1637? Or in what way did Shakespeare take into account the 'authenticity' of his writings when he wrote *Macbeth*? To restrict learning opportunities to mere 'authentic' learning situations is a hopeless deflation of education.

Furthermore, the evidence in historic practice is that 'authenticity' is in fact *not* required to increase the ability to learn or to retain information. Indeed, previous generations have survived sufficiently well in the absence of such authenticity. It could be argued (and often is) that this generation's educational needs are different from the educational needs of previous generations. To that end, this author agrees. However, it does not follow that the educational needs of this generation are better met in providing more 'authentic' learning opportunities.

### 3.2.3 The 'junglelike' brain

One final important question is: 'Why should the structure or the workings of the brain have any consequences for the structure and application of the curriculum, or for the nature and chosen forms of suitable assessment?' Moreover, why is it reasonable to supervene upon the chosen methods of providing learning opportunities for learners, the workings of the brain?

Robert Sylwester (a prominent neuroeducationalist who presented his ideas at CCEA's 'Mind Power 21: Educating Intellect and Emotion' conference) has this pseudoscientific answer, more suited to the realm of science fiction than to the world of rigorous scientific deliberation: a 'junglelike' brain is more likely to thrive in a 'junglelike' classroom, supported by a 'junglelike' curriculum.

> It suggests that a junglelike brain might thrive best in a junglelike classroom that includes many sensory, cultural, and problem layers that are closely related to the real-world environment in which we live – the environment that best stimulates the neural networks that are genetically tuned to it. … Educators might then view classroom misbehaviour as an ecological problem to be solved within the curriculum, rather than aberrant behaviour to be quashed. … Such a brain-based curriculum might resemble some current practice, but it might differ considerably from what schools are now doing. It's interesting to muse on such widely acclaimed developments as thematic curricula, cooperative learning, and portfolio assessment. … Is the appeal to educators that these approaches seem to be inherently right for a developing, junglelike brain, even though they require more professional effort and aren't nearly as economical and efficient as traditional forms? (Sylwester 1995: 23–4)

Sylwester's views are littered with conceptual blunders and misguided science, totally bereft of any form of rigour or scientific deliberation. What Sylwester means, precisely, by a 'junglelike brain' is far from clear.

Worse still, even if the notion of such a brain were to be established within science, it would remain a major piece of neuroscientific and educational work to draw a connective between a 'junglelike' brain and a 'junglelike' classroom. Indeed, why the nature of the brain should impact on the structure of the classroom and the curriculum is left unmentioned. This is as far as the 'explanation' goes.

If Sylwester had stopped there, he could have been excused as being honestly mistaken. However, he goes on to suggest that such neuroscientific understandings of education can assist with managing classroom misbehaviours, by treating them as 'ecological'. This is utterly inexcusable. There is *no evidence* to suggest *in any way* that a neuroscientific approach to classroom arrangement or curriculum structure can reduce classroom misbehaviours.

Further, he suggests that neuroscientific approaches to classroom and curriculum structure may assist teachers in viewing classroom misbehaviours in more manageable ways. It is irresponsible at best, and disgracefully opportunistic at worst, to offer teachers the hope of solving classroom misbehaviours via neuroscientific approaches to classroom and curriculum structure. In the attempt to set neuroscientific approaches up as the tonic for education to rid itself of its ailments, Sylwester has merely served to demonstrate the unforgivable bravado with which he puts forward unfounded ideas, as if they were established scientific facts. Even more unforgiveable was CCEA's adoption of such ideas in their curriculum reform; a reform which has already impacted and will continue to impact the education of many children, and the daily workings of many teachers.

It is noteworthy that such claims, on a broader scale, are not restricted to the domain of education. Indeed, in January 2015, neuroscience was introduced to the world of business, in the book *Neuroscience for Leadership*, littered with a host of claims of how one can further one's business acumen by learning how one's brain functions.

Rather worryingly, but also interestingly, the authors of this book (Swart, Chisholm and Brown) offer the secondary thesis that the brain can be compared to a business corporation, with the prefrontal cortex of the brain being viewed as the CEO of the brain.

This all sounds very much like Sylwester's claim of a junglelike brain demanding a junglelike curriculum, and, for that matter, Jensen's (2008) claim, outlined in Chapter 1, regarding the centrality of the brain inside the 'learning organisation' – the school. Setting aside the troubling neuroscientific errors which the book makes, outlined by neuroscientist Professor Steven Rose in his review of the book in *Times Higher Education* (12 February 2015) one can begin to appreciate how neuroscience has been invoked to support non-scientific claims with a degree of scientific credibility. Now, to be entirely fair, this is not an argument against the adoption of neuroscience within other disciplines, but it most certainly does serve as an argument against *bad* science in general being invoked to support pseudoscientific, unfounded inter-disciplinary projects. Rather ominously, Sylwester's neuroscientific thesis is such a project.

### 3.2.4 CCEA's responses

CCEA have subsequently offered attempts to salvage their position on the use of neuroscience in curriculum reform. The organization came under serious scrutiny for its appeal to neuroscientific evidence when Morrison (2006: 9) suggests, like this author, that CCEA have merely invoked neuroscience to 'bolster their case'. In response to this criticism, CCEA offered the following explanation:

> CCEA emphasizes, again, that neuroscience is not, and was not, the sole or prime foundation for the review of the Northern Ireland curriculum. The review was based on a raft of research, consultation and trialling to which neuroscience makes but one contribution. (CCEA 2006: 11)

CCEA are scathing of Morrison (2006), labelling him a neuroscientific sceptic, whereas they prefer to take a more 'middle ground' (CCEA 2006: 10). CCEA acknowledge John Hall's claims that 'neuroscience is still in its infancy' and that 'there is still much to be discovered, analysed and acted upon and that more work has to be carried out before any overarching theories integrating neuroscience, psychology and education can emerge' (CCEA 2006: 10–11). It is interesting then, that CCEA would have appealed to neuroscience *at all* in guiding their curriculum changes, given that – by their own admission – so much remains to be done before neuroscientific and educational ideas can be fused together. Perhaps time would have been better spent doing this *prior* to their curriculum change – at least some of which has been predicated on neuroscientific principles – instead of radically overhauling the curriculum structure on the basis of whimsical findings.

Gavin Boyd, CCEA's chief executive in 2003, disagrees with such caution, claiming that 'it seems foolish to wait until we are absolutely certain about everything, before we start to convey to our young people some of the basics about how the brain works and how this impacts on their learning' (Boyd, cited in Thompson and Maguire 2001: 2). Perhaps this demonstrates the laissez faire attitude taken to educational reform, where the CEO of the organization which is charged with implementing educational changes in Northern Ireland could suggest that sound and cogent evidence is no longer needed prior to implementing educational changes. Has education really become a 'let's see what happens' playground, in which changes are made and children are used as educational lab rats to test the latest educational fad? In the case of brain-based learning and neuroeducation, in particular in Northern Ireland, it seems so.

Moreover, Morrison (2006) is correct to critique CCEA's use of neuroscience, however little they claim to have appealed to it in their reforms. Indeed, in reading Morrison (2006), it is clear that it is not the *amount* of neuroscience that was used in CCEA's reform which is being criticized; rather, the criticisms were levelled at the *way* in which neuroscience was invoked to support curriculum reform in Northern Ireland. The fact that neuroscience was only one of many contributors to CCEA's case is, therefore, largely irrelevant, and offers little by the way of a defence of their thesis.

### 3.2.5 Contrary evidence from within

By way of example, it is clear to see precisely why educationalists must be careful in *what* research they call upon to assist them in making informed educational decisions. In CCEA (2006), the authors of the report cite Hall (2005: 20), who suggests that 'there is no "use it or lose it" phase' within learning. CCEA (2006) outline that in claiming this, Hall (2005) has identified an area of 'broad agreement' within the neuroscientific community.

Contrast this with the Royal Society's neuroscientific evidence which suggests that 'the brain changes constantly as a result of learning, and remains "plastic" throughout life. Neuroscience has shown that a skill changes the brain and that these changes revert when practice of the skill ceases. Hence *"use it or lose it" is an important principle for lifelong learning*' (The Royal Society 2011: 2, emphasis added).

So, who are the educationalists to believe? Hall (2005), who claims that, on the basis of broad neuroscientific agreement there is no such principle in learning as 'use it or lose it'; or the Royal Society (2011), and the collection of esteemed neuroscientific authors associated with their report, who claim that 'use it or lose it' is very much a part of successful learning?[2]

---

[2]  The neuroscientific literature is littered with conflicting ideas about explanations of learning. For example Geake (2005) suggests that reliable learning is best enforced by repetition (pp. 10–11). In contrast, as this author has already outlined, CCEA (2003) claims that learning is to be made 'authentic' and 'collaborative' to greater ensure successful learning. Both are allegedly supported by neuroscientific evidence.

## 3.3 The local problem, played out on the international stage

At this juncture, the reader would be excused for feeling like there is something of a strawman being created by the author, insomuch as the manifestation of the neuroscience–education collaboration has only been shown – somewhat curiously it must be said – to be confined only to a small region of the UK (Northern Ireland). If this was the only plausible case, then the neuroscience 'problem' – if indeed such a problem even exists – is hardly worth worrying about.

This local problem, however, now permeates a significant aspect of mainstream educational thinking. However, where the Northern Ireland situation clearly shows that neuroscience has been adopted in strict and overt curricular reform, the sea-change which is taking place elsewhere on the international stage is somewhat more tacit in nature.

In truth, brain-based approaches to education are nothing new; this much should already be clear from some of the preceding sections of this book, which show that brain-based education has been part of mainstream thinking and practice for almost two decades. Moving on from the early brain-based movement, neuroscience took centre stage and we then had the genesis of *Educational Neuroscience* (also referred to, as it has been in this book, as 'neuroeducation') and showed us that the early brain-based work was littered with blunders which ought to be disregarded, now referred to as 'neuromyths'.

On the back of this subtle shift, and with neuroscience jostling for position in this rapidly growing discipline, we now have a host of other sub-disciplines (which, in reality are in fact so rapidly expanding that they hardly qualify as 'sub' disciplines at all, rather they are disciplines in their own right) which take neuroscience to be the new informing science of education without reservation. Indeed, the birth of *Mind, Brain and Education* is one such sub-discipline which has been born out from the collaborative work between neuroscience, psychology and education, with an element of philosophy keeping everything 'in check'.

The brain simply is, without any question, taken to be the learning, thinking organ; and since education is interested in these concepts, it should, in principle, have much to learn from the findings taking place within neuroscience. And, since these findings are occurring on an almost regular basis, education simply *must* listen to neuroscience, lest it miss on this intellectual revolution which is taking place.

There is no doubt, in fact, that the growth in neuroscientific research now being conducted, and the vast amounts of money now being invested in said research, finds a precedent only in the growth of the medical sciences which took place almost a century ago. This rapid growth, therefore, and the expected implications of new findings due to advancements in technologies, gives rise to a new found hope that there is much to be learnt for any other disciplines which have an interest in the brain and its workings.

### 3.3.1 Rudimentary numbers, showing a culture change in educational research agendas

It is interesting to look at some rudimentary numbers, easily accessed by a quick Google search, to see how rapidly growing these disciplines are:

On Brain-based Education

- Google (general): 6.4 million results
- Google Books: 571, 000 results
- Google Scholarly Articles: 18, 600 results

On Educational Neuroscience/ Neuroscience and Education

- Google (general): 37.7 million results
- Google Books: 156, 000 results
- Google Scholarly Articles: 1.93 million results

On Mind, Brain and Education

- Google (general): 8 million results
- Google Books: 238, 000 results
- Google Scholarly Articles: 1.3 million results

To set these numbers in some kind of educational context, consider the same results for a commonly researched educational topic like 'male educational underachievement':

- Google (general): 697, 000 results
- Google Books: 8, 500 results
- Google Scholarly Articles: 30, 500 results

Or, consider the same numbers for perhaps the most researched topic within educational discourse 'education and poverty':

- Google (general): 150 million results
- Google Books: 673, 000 results
- Google Scholarly Articles: 2.4 million results

There is no doubt when one looks at these numbers (which *do not constitute research*, as opposed to a simple check on research popularity as accessible on Google), that neuroscience and education is fast becoming the game which everyone wants to be a part of. The charge of modern-day scientism has found education, and neuroscience is being offered as the science of learning.

The case in Northern Ireland, therefore, paradigmatic as is may be in the context of this book, may not be terribly far detached from the reality which will manifest in education within the next decade or less. As we have already seen in the Royal Society's (2011) endorsement of this collaboration, as well as endorsements from key players in the debate (Howard-Jones, Goswami, Frith, Geake, Blakemore, etc.), it seems entirely plausible that educational discourse will be influenced – perhaps even shaped – by neuroscience in the not-too-distant future.

Research centres dedicated to *Educational Neuroscience* have been convened at many universities across the globe, including Cambridge, Oxford, University College London, University of Bristol, Birkbeck, Harvard, Princeton, John Hopkins, King's College London, Yale and Stanford. Masters-level programmes leading onto PhDs are being offered at almost every major university one can fathom.

The neuroscience–education revolution is tacitly simmering at all levels within educational academia; which will naturally lead onto its eventual manifestation within primary and secondary schools all over the world, just as it already has in Northern Ireland. Indeed, as Zadina (2015) argues, educational neuroscience should be seen as *the* science which underpins curriculum reform (p. 73), teachers' professional development (pp. 73–4), and classroom practice and teaching strategies (pp. 73–4), despite neuromyths still plaguing such collaboration and practice:

> While brain research may not yet tell us *how* to teach *per se*, it does *inform* teaching, learning and school reform. We are *at the beginning of a new vision* [this author's italics] in which scientists, educators, and the hybrid Educational Neuroscientist can all work together toward school reform. (Ibid., p. 75, original emphasis)

This may indeed *seem* to be perfectly reasonable – perhaps even pragmatic – within educational policy reform. However, the Northern Ireland case study previously outlined should give us some *practical* cause for concern. Furthermore, this author is *not* simply making the age-old case that neuromyths still abound. Rather, there is also a philosophical case to be answered, given, as it will be shown later in the book, that the neuroscience–education collaboration is founded on questionable conceptual grounds.

Part Two

# The Philosophical Critique of Neuroeducation and Brain-based Learning: Mereology, Asymmetry and Irreducible Uncertainty

4

# The Mereological Fallacy

## 4.1 Introduction to Part 2

In this part of the book, the author will examine some of the philosophical and conceptual quandaries which befall the neuroscience movement in general, and the neuroscience–education collaboration in particular. This will be done by discussing three main principles of interest to this critique:

1. The mereological fallacy (Bennett and Hacker 2003)
2. The first-person/third-person asymmetry principle and the associated category error
3. Irreducible uncertainty of psychological predicates and attributes

## 4.2 Applying mereology to neuroscience: The Mereological Fallacy

Mereology is, broadly speaking, the study of whole/part relations. This section is dedicated to applying mereology to notions about the brain – as part – in relation to the human being, the whole.

This author argues that incoherent ascriptions of psychological predicates in general, and the predicates 'learn', 'think' and 'understand' in particular, to the brain in place of the human being are commonplace in the brain sciences, particularly and perhaps most prominently, within neuroscience (and educational considerations on, *mind, brain and education, brain-based learning* and *neuroeducation*). That is, predicates which can only be ascribed to the entire human being are often ascribed to the brain, in statements which consequently, it will be shown, transcend the bounds of sense.

The mereological argument, therefore, is used to refute the notion that the brain is a coherent domain to be taken as the *agent*[1] of psychological predicates in general, and the predicate 'learn' in particular. This author contests, in keeping with the philosophies of Wittgenstein, and subsequently (and most notably) of Peter Hacker, that such an ascription is not false as opposed to conceptually incoherent. Moreover, it will be argued pre-emptively that such incoherent ascriptions are not simply a case of deficiency in language, or mere metaphorical or metonymical talk, since the claims that are made are made in such a way that are intended to be literal. The particular interest of this section will be to demonstrate that all talk of the brain as the 'learning organ' is incoherent; a damning conclusion for the collaboration between neuroscience and education whose entire remit is predicated on such incoherencies.

### 4.2.1  A note from history: Phrenology and neuroscience

It is worth a short note also, that mereological-type arguments are not entirely new in the domain of brain science in general. Indeed, early pseudoscientific work in the late eighteenth and early nineteenth centuries by Franz Joseph Gall and Johann Gaspar Spurzheim led to the birth of phrenology, further developed by L. N. Fowler, which built upon work in the areas of organology and cranioscopy.

The idea was that the skull would be measured in order to ascertain the 'size' of different parts of the brain, and then certain cognitive and cogitative skills would be correlated with different parts of the brain, depending on skull, and thus brain size. In fact, a closer look at the work by Gall in phrenology, and his paper which was published in 1819 entitled *The Anatomy and Physiology of the Nervous System in General, and of the Brain in Particular, with Observations upon the possibility of ascertaining the several Intellectual and Moral Dispositions of Man and Animal, by the configuration of their Heads* shows that the foundations of phrenology were set as follows:

1. The Brain is the organ of the mind
2. The brain is not a homogeneous unity, but an aggregate of mental organs
   with specific functions

---

[1]  The argument against the brain being taken as the locus of psychological predicates is a separate argument found in the sections on *First-person/third-person asymmetry* and *The category error*. The distinction between the brain as the agent and the brain as the locus of psychological predicates is acknowledged by John Searle in *Neuroscience & Philosophy*, and is sufficiently dealt with later in the chapter.

3. The cerebral organs are topographically localized
4. Other things being equal, the relative size of any particular mental organ is indicative of the power or strength of that organ
5. Since the skull ossifies over the brain during infant development, external craniological means could be used to diagnose the internal states of the mental characters

(Lyons 2009: 53)

In truth, these principles, other than (4), could easily be found in most early neuroscience texts, and in reality, the only difference between the phrenological measurement (of the skull) and the neuroscientific measurement (the neural scan) is that where phrenologists sought to ascribe abilities to the brain on the basis of some *exterior* physical measurement (skull size), the neuroscientist transcribes such ascriptions to some *interior* physical measurement (of blood flow and oxygen levels in the brain). One wonders, then, why phrenology is disregarded as pseudoscience – despite its reasonable empirical success in establishing correlations, despite its conceptual quandaries – yet neuroscience escapes such criticisms, in light of the claim that it is a relatively 'young science'.

## 4.2.2 The Mereological Fallacy in neuroscience

Moving beyond this preamble, consider some examples of the Mereological Fallacy attributed to Bennett and Hacker (2003). Bennett and Hacker cite, in Bennett et al. (2007: 154–5), examples of neuroscientists who mistakenly attribute properties to the brain which are correctly attributed to the person, which this author cites in full to give an indication of the multitude of examples of the fallacy within neuroscience:

> *J Z Young*: We can regard all seeing as a continual search for the answers to questions posed by the brain. The signals from the retina constitute 'messages' conveying these answers. The brain then uses this information to construct a suitable hypothesis of what is there. (Source: Programs of the Brain: 119)
>
> *F Crick*: When the callosum is cut, the left hemisphere sees only the right half of the visual field … both hemispheres can hear what is being said … one half of the brain appears to be almost totally ignorant of what the other half saw. (Source: The Astonishing Hypothesis: 170)
>
> *S Zeki*: The brain's capacity to acquire knowledge, to abstract and to construct ideals. (Source: Royal Society B 354 (1999): 2054)

Bennett and Hacker (2003: 68–70) also cite other examples of such claims of the brain being capable of performing what would sensibly be understood as human activities:

*F Crick:* What you see is not what is *really* there; it is what your brain *believes* is there. … Your brain makes the best interpretation it can according to its previous experience and the limited and ambiguous information provided by your eyes. … The brain combines the information provided by the many distinct features of the visual scene (aspects of shape, colour, movement, etc.) and settles on the most plausible interpretation of all these various clues taken together. … What the brain has to build up is a many-levelled interpretation of the visual scene. … [Filling-in] allows the brain to guess a complete picture from only partial information – a very useful ability. (Source: The Astonishing Hypothesis: 30, 32f., 57)

*G Edelman:* [The brain can] 'categorize, discriminate, and recombine the various brain activities occurring in different kinds of global mappings' [and it] 'recursively relates semantic to phonological sequences and then generates syntactic correspondences, not from perplexing rules, but by treating rules developing in memory as objects for conceptual manipulation.' (Source: Bright Air, Brilliant Fire – On the Matter of the Mind: 109f., 130)

*C Blakemore:* We seem driven to say that such neurons [as respond in a highly specific manner to, e.g., line orientation] have knowledge. They have intelligence, for they are able to estimate the probability of outside events – events that are important to the animal in question. And the brain gains its knowledge by a process analogous to the inductive reasoning of the classical scientific method. Neurons present arguments to the brain based on the specific feature that they detect, arguments on which the brain constructs its hypothesis of perception. (Source: Mechanics of the Mind: 91)

*A Damasio:* Our brains can often decide well, in seconds, or minutes, depending on the time frame we set as appropriate for the goal we want to achieve, and if they can do so, they must do the marvellous job with more than just pure reason. (Source: Descartes's Error – Emotion, Reason and the Human Brain: 173)

*B Libet:* The brain 'decides' to initiate or, at least, to prepare to initiate the act before there is any reportable subjective awareness that such a decision has taken place. (Source: Unconscious cerebral initiative and the role of conscious will in voluntary action, *Behavioural and Brain Sciences, 8*: 536)

*J Frisby:* There must be a symbolic description in the brain of the outside world, a description cast in symbols which stands for various aspects of the world of which the sight makes us aware. (Source: Seeing: Illusion, Brain and Mind: 8f)

*R Gregory:* [Seeing is] probably the most sophisticated of all the brain's activities: calling upon its stores of memory data; requiring subtle classifications, comparisons and logical decisions for sensory data to become perception. (Source: The confounded eye, cited from *Illusion in Nature and Art*: 50)

*D Marr:* Our brains must somehow be capable of representing … information. … The study of vision must therefore include … also an inquiry into the nature of the internal representations by which we capture this information and make it available as a basis for decisions about our thoughts and actions. (Source: Vision, a Conceptual Investigation into the Human Representation and Processing of Visual Information: 3)

*P Johnson-Laird:* [The brain] has access to a partial model of its own capabilities' and '[has] the recursive machinery to embed models within models' and '[consciousness] is the property of a class of parallel algorithms. (Source: How could consciousness arise from the computation of the brain?, cited from *Mindwaves*: 257)

These all offend against the 'mereological fallacy'. Upon closer inspection of all of these claims, it is possible to see what each has in common with the others: namely, that what is normally and sensibly posited of the entire human being, is now being posited of the brain *as part of* the human being.

The problem is also rife within educational discourse. For example, as it has been shown, in Northern Ireland, the CCEA have adopted neuroscientific principles as part of the underpinning rationale for curriculum reform. Recall how CCEA's chief executive in 2003, Gavin Boyd, claims, 'It seems foolish to wait until we are absolutely certain about everything, before we start to convey to our young people some of the basics about how the brain works and how this impacts on their learning' (Boyd, cited in Thompson and Maguire 2001: 2). The 'basics' which Mr Boyd was citing included the claim that 'thought is filtered through the emotional part of the brain first' (CCEA 2003b: 22). Moreover, CCEA cite evidence from neuroeducationalist Robert Sylwester, who claims that a 'junglelike' brain is more likely to thrive in a 'junglelike' classroom, supported by a 'junglelike' curriculum (Sylwester 1995: 23–4).

The salient point is, however, that these claims all offend against the so-called mereological fallacy. It is well understood what it is for a human being to: 'pose questions', 'use information', 'see', 'hear', 'be ignorant of' and 'develop knowledge', but it is doubtful if the day will come when it can be demonstrated in the laboratory that a brain can do any of these things. Moreover, what precisely is meant by claims that the brain is 'junglelike' is far from clear, and not fitting of any form of scientific rigour.

To mistakenly attribute properties to the brain which are, in fact, properties of the human being is to fall prey to the mereological fallacy. As Bennett and Hacker claim:

> Psychological predicates are predicates that apply essentially to the whole living animal, not to its parts. It is not the eye (let alone the brain) that sees, but *we* see *with* our eyes (and we do not see with our brains, although without a brain functioning normally in respect of the visual system, we would not see). So, too, it is not the ear that hears, but the animal whose ear it is. The organs of an animal are part of the animal, and psychological predicates are ascribable to the whole animal, not its constituent parts. Mereology is the logic of part/whole relations. (Bennett and Hacker 2003: 72–3)

By talking of the brain as 'learning', 'thinking', 'believing' or 'understanding' the neuroscientist is mistakenly reducing psychological predicates (ascribable only to the entire human being) to brain states (ascribable to the brain). Those who offer seminars[2] and in-service training in 'brain-based learning' often refer to brains 'thinking', 'knowing', 'believing', 'deciding', 'seeing an image of a cube', 'reasoning', 'learning' and so on. Bennett and Hacker, however, contest:

> We know what it is for human beings to experience things, to see things, to know or believe things, to make decisions. … But do we know what it is for *a brain* to see … for *a brain* to have experiences, to know or believe something? Do we have any conception of what it would be like for *a brain* to make a decision? … These are all attributes of human beings. Is it a new discovery that brains also engage in such human activities? (Bennett and Hacker 2003: 70)

Bennett and Hacker (2003: 70–1) conclude by giving the neuroscientist three options as to what precisely they might mean when they become so apparently loose with their language about the brain's abilities:

1. Is it a new discovery that the brain can engage in human activities? OR;
2. Is it linguistic innovation for the cognitive sciences to extend ordinary language to a specialist use? OR;
3. Is it conceptual confusion to posit that the brain can do things normally and sensibly ascribed to a human being?

---

[2] A particularly good example of which is *Fingerprint Learning*, a company which offers brain training for teachers and pupils in Northern Ireland and the UK.

### 4.2.3 Empiricism vs. Conceptualism

Embedded into the previous discussions is a clear distinction between empirically focused philosophers and scientists, and their conceptually focused counterparts. The point of contention between these two camps seems to be where one defines one's starting point in one's scientific endeavours and investigations.

Philosophers like Hacker, for example, contest that 'empirical research cannot resolve any philosophical problems' (Bennett and Hacker 2003: 414), whereas philosophers such as Dennett counter-argue 'well of course not; empirical research does not *solve* them, it *informs* them and sometimes *adjusts* or *revises* them' (Bennett et al. 2007: 80).

Hacker's position, like that of Ryle and Wittgenstein, is governed by a strict adherence to so-called 'ordinary language' philosophy, which Dennett in particular regards as a 'bluff' (Bennett et al. 2007: 83). In other words, what Hacker in particular might contest to be a transgression on the bounds of sense, Dennett and his followers would argue is a simple extension of language for innovative purposes, granted under the linguistic licence of science.

To clarify the point, it is worthwhile to return to Wittgenstein's claim that 'philosophical problems arise when language *goes on holiday*' (*PI*, §38). Furthermore, Wittgenstein also claims, 'The problems are solved, not by coming up with new discoveries, but by assembling what we have long been familiar with', and concludes, 'Philosophy is a struggle against the bewitchment of our understanding by the resources of our language' (*PI*, §309).

These realizations are important in unpacking Wittgenstein's claims about the resemblance of a human being, and how we can meaningfully ascribe the predicates of a human being to that which closely resembles a human being. In the case of neuroscience, Bennett and Hacker warn that 'it is the task of the conceptual critic to identify ... transgressions on the bounds of sense', and whenever such transgressions are spotted, the critic must condemn the wayward scientist for his abuse of language (2003: 6).

Dennett argues that all scientific questions have the power to inform and transform the scientific edifice. This author however, like Bennett and Hacker, Ryle and Wittgenstein, for example, warns that science *cannot* and indeed *ought not to* be informed by an incoherent conceptual basis. Perhaps, as the eminent physicist and scientific philosopher, Thomas Kuhn suggested, science should have a sound conceptual basis, inside which sensible questions can be posed, intelligible empirical findings can be made, and descriptions and theories can be formed, reformed and evolved through time. None of this is possible, however, if the conceptual foundations of one's investigations are flawed.

### 4.2.4  Incoherent, NOT false

The profound error captured in the mereological fallacy must be carefully acknowledged. When the neuroscientific claim that brains process information is called into question, this does not render valid the assertion that brains, in fact, *do not* process information. That is, in rejection of psychological predicates being ascribable to the brain, *it does not follow* that the negation is in fact the case. In essence, it is not the case that brains *do not* think, learn and hypothesize; instead, as Bennett and Hacker (2003) claim, *it makes no sense* to utter such words.

The conclusion is, therefore, that it is not false that brains carry out the activities which would normally be ascribed to a human being, as opposed to *it making no sense* to talk of such things. In this way, we return to Wittgenstein's original philosophical position, insomuch as the incoherent ascription of psychological predicates to the brain in place of the human being is senseless because of a transgression on the bounds of sense. It makes no sense to talk of brains doing any of these things, since it is unclear what it would even mean for a brain to perform these tasks. That is, the sentence 'the brain learns' (and other such similar claims) is a sentence whose construction arises from the juxtaposition of terms which cannot coherently be placed alongside one another, without infringing on the boundaries of sense. This is why Wittgenstein spoke of such instances as examples of when 'language goes on holiday' (*PI*, §38); because to talk of the brain 'learning', for example, is to use the word 'learn' in a manner other than what its use permits. And the result is nonsense; not falsehood, but nonsense.

Similarly, the claim that the brain does not learn, is senseless, since one can only ascribe a property *or its negation* to an object, once one has established that such an ascription is even possible, without deviating away from sense. For example, to say that a lamp is not happy does not render valid the assertion that the lamp is sad; rather, that lamps cannot be spoken of sensibly as being capable of being 'happy' or 'sad'. A clock cannot be said to tell the time with 'passion'; but this does not suggest that the clock is 'passionless'. Instead, all talk of clocks being either with or without 'passion' is dismissed as nonsense. Similarly, to talk of the brain 'learning' or 'thinking' is a failure to grasp the sense of one's words. It is not that brains do not learn or think; rather that we have lost a grip on sense when we utter such words.

To the neuroscientist who talks of the brain 'learning', 'thinking' and 'hypothesizing', one might reply: 'And what precisely does that look like?' As Bennett and Hacker summarize:

> It is our contention that this application of psychological predicates to the brain *makes no sense*. It is not that as a matter of fact brains do not think, hypothesise and decide, see and hear, ask and answer questions; rather, it makes no sense to ascribe such predicates *or their negations* to the brain. The brain neither sees, nor is it blind – just as sticks and stones are not awake, but they are not asleep either. (Bennett and Hacker 2003: 72)

Advocates of brain-based learning see the brain as the agent[3] of thought; that is, the brain is the learning organ. Wittgenstein, while accepting that without a properly functioning brain one could not think, nevertheless teaches that thinking is done by the whole person, and not the brain.

When a teacher asks a pupil what she thinks, the pupil *expresses* her thoughts in language. Were it not for the pupil's language skills, the teacher could not ascribe thoughts to her. Since brains aren't language-using entities, how can it *make sense* to ascribe thoughts to a brain? That is, since the brain is not the human being, and does not resemble the human being (insomuch that it is not a language-using entity), it follows that the properties or abilities of the human being cannot be ascribed to the brain, which is itself only a part of the human being. To conclude:

> Our point, then, is a conceptual one. It makes no sense to ascribe psychological predicates (or their negations) to the brain, save metaphorically or metonymically. The resultant combination of words does not say anything that is false; rather it says nothing at all, for it lacks sense. Psychological predicates are predicates that apply essentially to the whole living animal, not to its parts. (Bennett and Hacker 2003: 72)

### 4.2.5  Neural imaging: PET and fMRI scans

Neuroscientists may protest that the brain's ability to make connections while it (the brain) is thinking, learning or hypothesizing is visible from PET or fMRI images of the brain which are a staple of the cornucopia of neuroscientific methods. Bennett and Hacker (2003: 83–4) reject this notion:

> What we can do is correlate a person's thinking of this or that with localized brain activity detected by PET or fMRI. But this does not show that the brain

---

[3]  Some say the brain is not the agent, but instead is the locus of psychological predicates. This will be examined in Chapter 5.

is thinking, reflecting or ruminating; it shows that such-and-such parts of a person's cortex are active when the *person* is thinking, reflecting or ruminating.

That is, what might appear on a neural image to be the brain thinking or learning is simply the brain activity which can be correlated to the person thinking or learning whose brain it is. PET and fMRI scans, therefore do not show the brain thinking or learning, as opposed to showing the areas of the brain which are used while the person is thinking or learning and the areas of the brain in which activity occurs while the person is doing these things.

Moreover, as Bennett and Hacker argue, the measurement of such activity and its correlation to the human being's behaviours are predicated on the criterial understanding of what it is to display 'learning' or 'thinking' behaviours: '[The correlation] presupposes the *concept* of thinking, *as determined by the behavioural criteria that warrant ascription of thought to a living being*' (2003: 70–1). That is, were it not possible to identify someone who is thinking or learning in the 'outward' behavioural sense, and if it were not possible to ascribe thinking and learning to the human being on the basis of their demonstrable behaviour, it would not be possible to establish any correlation between brain scans or neural images and the behaviour.

In this way, the correlations drawn between brain activity, the areas of the brain which are active during certain behaviours and the outward behaviours themselves is *dependent upon* there being criteria for the accompanying behaviour in question. The criteria for the behaviour which is correlated to the brain activity are, therefore, pre-supposed in any neural image which goes in search of the correlated brain activity to the behaviour. If there were no such thing as 'learning-behaviour', and if there were no criteria upon which 'learning-behaviour' could be ascribed to the human being, then there would be no correlations to be drawn between brain scans, the brain activity which they show and the behaviour to which they are correlated.

To emphasize Bennett and Hacker's (2003) point, it is useful to understand precisely what these neural imaging techniques show in terms of brain activity, and also to outline the precise details of the techniques used to do it. In Howard-Jones (2010), Usha Goswami outlines what the fMRI technique entails:

> The fMRI technique measures changes in blood flow in the brain, which can take approximately 6-8 seconds to reach a maximum value (i.e. maximum activity will be measurable 6-8 seconds after reading a particular word). fMRI works by measuring the magnetic resonance signal generated by the protons of water molecules in brain cells, generating a BOLD (blood oxygenation level

dependent) response. The fMRI method is excellent for localisation of function, but because changes in brain activity are summated over time, it cannot provide information about the sequence in which neural networks become engaged during the act [of reading]. (Goswami 2010: 19)

And of the PET technique, Goswami outlines that 'radioactive tracers are injected into the bloodstream and provide an index of brain metabolism' (ibid.).

Of the same techniques, Wright (2010: 5) outlines, quoted at length:

Functional magnetic resonance imaging can show which part of the brain is active, or functioning, in response to the patient performing a given task, by recording the movement of blood flow. All atoms and molecules have magnetic resonance, emitting tiny radio wave signals with movement, because they contain protons. Different molecules have different magnetic resonance and two components of blood are tracked to observe brain activity.

Haemoglobin in the blood carries oxygen; oxyhaemoglobin, around the brain and when it is used up, it becomes desoxyhaemoglobin. Where the oxygen is being 'used up' shows the site of activity in the brain. The picture is made by monitoring the ratio of the tiny wave frequencies between these two states whilst the patient carries out the task, e.g. tapping a finger, which highlights the area of the brain functioning to carry out this task.

Similarly, there is a detailed outline of what PET scans entail given by Wright (2010: 6), again cited at length:

Positron emission tomography scanning produces a three-dimensional image of functional processes in the brain (not just the structure). PET is a nuclear medicine imaging technique which requires the patient to receive a small injection of radio-active material (a sugar-tracer; fluorodeoxyglugose), into the bloodstream. The radio-active material causes the production of gamma-rays. These are a form of electromagnetic radiation like X-rays, but of higher energy. The radio-active material is transported around the body and into the brain. A ring of detectors outside the head is used to detect pairs of gamma rays emitted indirectly by the positron-emitting radionuclide (tracer), in each part of the brain under examination.

The areas of the brain that command the greater volumes of blood produce the most gamma-rays, and it is these areas that are computed and displayed by the PET scan. As the tracer decays, there is a point when gamma photons are emitted almost opposite to each other. … The timing of this event is detected and will ultimately improve the detail of the image. The system not only identifies the activated area of the brain, but also measures the degree of activity.

Other than a minor digression from the facts by Wright (2010), in talking about 'the area of the brain functioning to carry out this task' – which is another example of ascribing abilities to the brain (carrying out tasks) which can only be sensibly ascribed to the human being – these definitions and descriptions of PET and fMRI techniques respectively outline nothing other than the fact that these neural imaging methods can simply identify brain activity represented by blood flow in certain areas of the brain.

Nowhere in these generally accepted descriptions is there evidence to support the claim that what the blood flow represents is actual *learning* or *thinking*. That is, these sophisticated neural imaging techniques show nothing more than blood flow and brain activity; and since there is no comprehension of learning or thinking as mere blood flow or brain activity, any suggestion that these things are actual learning or thinking as brain activities is a flawed induction.

There are no empirical difficulties with drawing correlations between brain activity and the behaviours which seem to be connected to them. However, such an admission *is not the same* as suggesting that the blood flow or brain activity *is* the learning or thinking itself. The value of neural imaging, therefore is to draw correlations between brain activity and outward behaviour; a model which is *dependent* on observable behaviour (and the criteria for identifying such behaviour) as well as neural imaging techniques.

For the moment, however, it suffices to say that all that can be learnt from the most sophisticated neural imaging techniques such as PET and fMRI is that there is activity in certain areas of the brain which can be probabilistically correlated to certain behaviours *while* the living human being does them. And to emphasize once more, such correlations *do not show* that the brain learns, thinks or hypothesizes, as opposed to showing the activity in the brain while the human being, whose brain it is, does these things.

### 4.2.6  Is it all literal? Closing the metaphorical and metonymical loops

It is important to note that the mereological fallacy is not simply a consequence of a lack of suitable language at the disposal of the neuroscientist. When there are claims of brains learning, thinking and understanding, there are three options open to outline the nature of such claims:

1. The intentional predicates in question, when ascribed to the brain, are extended versions of the intentional predicates that are normally ascribed to the human being;

2. Ascribing intentional predicates to the brain is metaphorical;
3. Ascribing intentional predicates to the brain is linguistic innovation, where the terms in question are used in a technical sense.[4]

If none of these choices can be made, then the jump from ascribing intentional predicates to the human being, to ascribing these predicates to the brain is unjustified, meaning the conceptual foundations of neuroscience require reform, and the purported common interest between neuroscience and education is predicated on questionable assumptions.

In the case of choosing option (1), when the everyday versions of intentional predicates are extended to the brain, this process of extension generates conceptual confusion in distinguishing the old use from the new extended use. In this way, the extension of intentional predicates to the brain gives rise to the question of what it *means* for the brain to be an intentional entity. And since no answer to this question has been provided, and no experiments have been done to *show* what it means for a brain to learn or think in the 'extended' sense of these words, it follows that to extend these ascriptions of intentional terms to the brain in place of the human being is unjustified, leading only to the mereological fallacy, as previously outlined.

Other philosophers and neuroscientists opt for option (2), that this talk of the 'learning, thinking, seeing, hypothesizing' brain is the talk of a mere metaphor. This is the loop which this author will show must be closed off. That is, the option of suggesting that this is a metaphorical extension of the intentional predicates in question is not open to the neuroscientist, and this is because of the way in which they use the terms. For example, when organizations like CCEA invoke neuroscience to support the claim that 'thought is filtered through the emotional part of the brain first' (CCEA 2003b: 22) they are using the word 'thought' in the way it is usually used in everyday psychological vocabulary. No effort is made to distinguish between what 'thought' is normally understood to mean, and what it means in CCEA's use of the word in this case, to mean a traceable 'object' which *flows* through the emotional part of the brain.

Similarly, when prominent neuroscientists talk of the brain's seeing and hearing, thinking and understanding, being ignorant of or learning, acquiring knowledge and forming and testing hypotheses, they use these terms in such a way which is an extended version of the use of the corresponding everyday terms.

---

[4] It is worth a note that Searle outlines in (Bennett et al. 2007: 108–12) that there are also other arguments which are not accounted for in Bennett and Hacker's mereological fallacy. They will be considered in Chapter 6.

In cases such as Dennett's (1987) extension of such terms to apply the intentional stance to the brain and its parts, the extended version – or the 'attenuated' version, as Dennett called it – stripped of many of its everyday connotations, is often used in such a way that it becomes impossible to distinguish between the old use and the new.

Such extensions are not metaphorical; that is, when authors like Dennett claim that the behaviours of the brain (and its parts) are 'strikingly like' the behaviours of the whole human being, the extension of words which define human behaviours to be used to describe and outline brain behaviours is a literal extension, not a metaphorical one. The extension may have 'stripped' the original word of many of its everyday characteristics, but the fact remains that the extension is a literal extension of intentional predicates and human behaviours as opposed to a metaphorical one.

In the case of vying for option (3), it is clear that there has been no effort to explain what is meant by using these terms in a technical sense. Moreover, if these terms are being used in a technical manner, those who choose to use them in this way ought to be careful not to forget that they are being used technically and *not ordinarily*. Too many neuroscientists, like those cited in §4.22 of this book, predicate research on the 'technical' use of such terms, and then forget – somewhere along the line – that these terms were not intended to be used in their ordinary context.

In any case, even if the use of such terms for supposedly innovative language in the pursuit of explanations and hypotheses is not entirely literal, or if the terms are being used in some technical sense, there is a component of each of these intentional terms which is a residue from the ordinary use of the word; enough of a residue to cause confusion, and give rise to the notion that these terms *are* being used literally, and/or ordinarily.

If the neuroscientist talks of a brain 'learning' or 'thinking' in a way which is contrary to or different from the human being doing these things (even in the 'technical' sense, whatever this might mean), they had better make clear precisely what it *means* for a brain to perform these roles. That is, it is clear what is meant by the human being learning, thinking and understanding, seeing and hearing and so on, and so if there is some difference between what it means for a human being to do these things, as opposed to what it means for the brain to do these things, perhaps it is about time this difference was articulated.

Indeed, if such a contrast could be put into words, perhaps Bennett and Hacker's argument in *Philosophical Foundations of Neuroscience* could be dispensed with as one major philosophical oversight. Perhaps the most ominous fact for neuroscience as a discipline is that such an articulation has never occurred. That is, it has never been explicitly and coherently stated what it *means*, precisely, to talk of the brain as the subject of intentional predicates or to outline what faculties the brain has, to be considered a coherent domain as an object that can decide to take action, to act, and to reflect on its actions, on the basis of its intentional faculties. The ordinary use of these intentional predicates – predicates which can and are usually ascribed to the entire human being; and it is well defined what it is for the human being to be considered an intentional being – has been extended, without the due diligence required to account for the incoherencies with uttering such combinations of words.

In essence, therefore, the defence that this is all just a matter of semantics, that reference to the brain thinking is little more than a *façon de parler*, seems not to be open to neuroscience. To make their case, CCEA, Sylwester, Crick, Young, Edelman, Dennett and the many others cited in previous sections who have fallen foul of the mereological fallacy in the claims that they make must be claiming that intentional predicates *can coherently* be ascribed to the brain or to its parts (e.g. neurons and the like), as entities capable of intentionality. That is, the 'learning, thinking, seeing, intentional' brain is not the talk of mere metaphor.

All the avenues are therefore closed, and the neuroscientist has nowhere to turn. All talk of the 'learning, thinking, deciding, understanding, hypothesizing' brain is non-metaphorical, incoherent pseudoscience, with weak foundations and substandard conceptual underpinnings.

### 4.2.7 Tying the mereological fallacy to neuroeducation and Brain-based learning

Brain-based learning and neuroeducation are predicated, at least in part, on principles which assume the brain as the subject of the intentional predicates, such as learn, think and understand. It is held within the scientific underpinnings of these educational 'theories' that the brain is the learning organ. Since it has been shown that the brain is in fact not a coherent domain to be the agent or subject of learning, thinking and understanding (and other such intentional predicates), it follows that brain-based learning and neuroeducation are predicated on conceptual blunders and incoherencies. These 'theories' are therefore fundamentally flawed.

It has been demonstrated how brain-based learning approaches – in particular neuroeducation – are guilty of the mereological fallacy, in that these approaches advocate the incoherent ascription of psychological predicates to the brain, which forms only a *part* of the object – the human being – to which these psychological predicates can be coherently ascribed. Failing to understand the mereological nature of the brain in relation to the entire human being leads to such an incoherent ascription.

The brain is a mere part of the entire human being; and to talk of the brain as the learning, thinking, understanding organ is guilty of committing the mereological fallacy. To ascribe learning, thinking and understanding to the brain as the organ responsible for these abilities is incoherent, not false. The notion of collaboration between neuroscience and education is therefore predicated on a conceptual blunder on the basis of the fact that neuroscientists apply the attributes of interest to education to the brain in place of the human being in an incoherent manner, and so the adaptation of neuroeducational principles in schools by teachers would be to commit the same error. What this seems to point towards is that the concepts and phenomena of neuroscience *in its current form* are not amenable with the concepts and phenomena of education.

Furthermore, since the notion that the brain is the domain of learning, thinking and understanding has been undermined, it follows that the value of neuroscience to education in any potential collaboration is also undermined, in particular on the grounds that the proposed areas of 'overlap' between neuroscientific and educational discourses are precisely these areas of conceptual incoherence.

To be clear, *this does not imply that neuroscience has nothing to offer to education*. There is, in fact, little doubt that there is a great deal to learn from the brain sciences as to how the *human being* learns, thinks and understands things. Indeed, a fusion between the neurosciences, psychology and education may well provide such insights. Nevertheless, until such potential collaboration is founded on something other than the incoherent notion that it is the brain which learns, or for that matter, that learning takes place *inside* the brain, then such collaboration is destined only to fail. Indeed, if education is to learn anything from neuroscience at this primitive stage, it is precisely that the brain is a mere part of the learning organ which is the *human being* whose brain it is.

The brain is neither knowledgeable, nor ignorant; it does not understand, but neither does it misunderstand; it is not intelligent, but it is not unintelligent either, for these are all things predicated only of the entire human being. A fully functioning brain is used in the human being becoming enlightened, or becoming more intelligent; but it is not the brain which is furnished with these faculties, as opposed to the human being whose brain it is. Education, therefore, would do better to focus on how the *human being* can be given a greater propensity to learn, to become more knowledgeable or to understand, for it is the human being who has these abilities.

# First-Person/Third-Person Asymmetry

## 5.1 Interlude

In Chapter 4, it was shown how the conceptual problems captured in the mereological fallacy seem to preclude the neuroscientific premise that the brain can be considered as the domain of the coherent ascription of psychological predicates that would normally and intelligibly be ascribed to the entire human being. This is a rejection of the notion that the brain can be considered as the *agent* of psychological attributes. However, such an argument does not preclude it being possible that the brain can reasonably be considered as the *locus* of said predicates and attributes. So, where it might be wrong to say 'the brain learns, and such learning is shown on the neural scanner', it does not follow automatically that it is wrong also to say 'learning takes place *inside* the brain, and this is seen on the neural scanner when the learning takes place'. Further consideration of this second case is required.

## 5.2 First-person/third-person Asymmetry

It seems intuitive to suggest – and many do – that inner brain activity *causes* observable behaviour. According to this view of things, learning starts in the brain and manifests in behaviours. The brain is *where* learning, thinking, etc., occur, and the evidence for this, it is argued, can be seen when parts of the neural image 'light up' when the person learns, thinks or ponders that things are thus-and-so. The seemingly natural conclusion is that the brain is the *location* of psychological attributes, and therefore what the neural scan shows simply *is* a representation of the phenomenon in question.

The natural extension of this premise is that the psychological attributes of interest to education can be investigated using neuroscientific methods.

Despite the quandaries of the mereological fallacy which reject the notion of the brain as the agent of such attributes, it is perfectly fine – it is contended – to posit that the brain is the *location* of such phenomena, and so our investigations should be focused on establishing what factors play a role in making the brain a more likely place for learning, thinking and understanding to take place *inside*. Such a model contests that although it might not be the brain which actually 'does' the learning, etc., it might still be the *location where* such activity occurs and that such activity is simply neural/brain activity (cf. Searle in Bennett et al. 2007: 111).

Learning and thinking, then, simply *are* brain activity. Neurophilosopher, Searle, for example, claims that, 'with the development of fMRI and other imaging techniques, we are getting closer to being able to say *exactly where* in my brain the thoughts occur' (Bennett et al. 2007: 109, emphasis added).

This is a troubling argument, in the main due to the so-called first-person/ third-person asymmetry principle, which will be outlined over the course of this chapter.

## 5.2.1 The asymmetry principle

Psychological and educational predicates are governed by the 'asymmetry principle', succinctly described by Harré and Tissaw:

> In making a first-person statement I am making an avowal. I am expressing how it is with me, sincerely or insincerely. In making a third-person statement about somebody else's feelings I am describing that person's feelings correctly or incorrectly. In the first case I need no evidence. In the second case I must go on the signs I see or hear. These distinctions between the grammar of first-person expressive talk and third-person descriptive talk can be categorised as the asymmetry principle. (Harré and Tissaw 2005: 190)

Furthermore, psychological verbs such as learning, thinking and understanding are to be understood and used with reference to the first-person/third-person asymmetry which governs them and their use:

> The truth is that psychological terms are not the names of psychological properties, acts or activities *in the sense which* 'red' can be said to be the name of a colour property, 'to wave' can be said to be the name of an act, and 'to dig the garden' can be said to be the name of an activity. To be sure, it is not wrong to say that 'pain' is the name of a sensation, or that 'anger' is the name of an emotion. But this masks deep logical differences between being the name of an act or activity

and being the name of a so-called mental act or mental activity. It suggests a similarity in kind of expressions where there are important logical differences, obscuring their profoundly differently uses. (Bennett and Hacker 2003: 100–1)

This realization then leads to the asymmetry principle of first- and third-person uses of such psychological verbs:

> Psychological predicates are attached, for example, to personal names, pronouns or definite descriptions. There is an asymmetry between the first-person, present tense use of these words and the corresponding third-person use. Indeed, this much is recognized even by the received conception, for, on the favoured view, we apply these predicates to ourselves on the basis of introspection or inner perception, whereas we apply them to others on the basis of the non-logical evidence of their behaviour. But this misconstrues the nature of asymmetry, which is rather that we apply (a subset of) them to ourselves without any 'basis' or evidential grounds at all, and apply them to others on the basis of behavioural criteria – which constitute logically good evidence.
>
> To learn the means of those psychological predicates that are of concern to us, one must master both their first-person and their third-person uses. For these are different facets of one and the same concept. (Bennett and Hacker 2003: 101)

Furthermore, Wittgenstein outlines the principle as follows:

> Psychological verbs [are] characterized by the fact that the third person present is to be identified by observation, the first person not. Sentences in the third person of the present: information. In the first person present expression. (*RPP II*, §63)

Also, Wisdom (1967) opines that the asymmetry principle which governs psychological (and educational) predicates is not a matter for regret:

> The asymmetrical logic of statements about the mind is a feature of them without which they would not be statements about the mind, and that they have this feature is no more a subject suitable for regret than the fact that lines, if truly parallel don't meet. (Wisdom 1967: 361)

Similarly Suter adds to Wisdom's view that the asymmetrical nature of psychological predicates is simply a feature of them in no need of regret, when he claims: 'This asymmetry in the use of psychological and mental predicates – between the first-person present-tense and second- and third-person present-tense – we may take as one of the special features of the mental' Suter (1989: 152–3).

So, the asymmetry principle in how psychological predicates are ascribed to the first and third persons is noticeable in the nature of each of these ascriptions:

to the first-person (to oneself) *without* criteria, and to the third-person (someone else) *with* behavioural criteria, which are logically sound grounds for their ascription, that is, publically 'agreed' behavioural criteria.

It is worth noting that the asymmetry principle is seen as one of the major cornerstones of Wittgenstein's argument against Cartesianism. Indeed, Wittgenstein's account of first-person/third-person asymmetry helps to sidestep the philosophical complications which are induced by the Cartesian doctrine:

> Cartesianism takes this first person-certainty to be a matter of infallible knowledge, owing to a person's having direct access to his or her thoughts and feelings by means of introspection. For this perspective, first person utterances ... should be considered to be knowledge claims based on inward observations, or reports of the things someone perceives inside. Our access to other minds, by contrast, is supposed to be indirect, and Cartesianism accordingly takes third person statements ... to amount to educated guesses at best. Based on an inventory of external clues rather than an inspection of a person's inner itself, third person statements can supposedly never shake off their speculative status. (Bax 2011: 40–1)

Many of the details of the preceding exert will be examined in greater detail in subsequent sections and chapters. For the moment, however, it is clear that Wittgenstein's development of the concept of first-person/third-person asymmetry is often invoked as a refutation – at least in part – of the core precepts of Cartesian thinking.

> Wittgenstein shows that within a Cartesian framework – no matter how well equipped it seems for capturing the certainty distinctive of the first and the uncertainty distinctive of the third person utterances – both types of statement are misrepresented, both in terms of the kind of access purportedly enabling them, and in terms of the epistemological status granted to them. (Bax 2011: 41)

For the moment, let it be clear that this is not the final application of the asymmetry principle. Many of the ideas outlined above will be developed in significantly more detail later.

## 5.2.2 The symmetry principle

In contrast, the way in which physical attributes are ascribed to the first- and third-persons follows a *symmetry principle*, insomuch that it is done in the same way for oneself as it is for another. To say that someone is such-and-such a height, for example, is done with no distinction between the first- and third-person cases. There are no criteria for the third-person ascriptions of height that are different for the way in which one would ascribe a height to oneself. In this way, physical attributes, for example, follow the symmetry principle of first-person/third-person ascriptions.

### 5.2.3 Not a something, but not a nothing either!

Supporters of the neuroscientific and brain-based learning doctrines fail to grasp the asymmetry principle of ascription when they posit that, by using sophisticated neural imaging techniques, they can attain access to the previously inaccessible hidden, inner realm, about which the teacher knew nothing.

The tendency is to spurn behaviourism (where mental states could be explained wholly in terms of 'bare' behaviour) for a crypto-version of Cartesian dualism (where the functioning of the brain is separable from behaviour).

But, in keeping with Wittgenstein, it is clear that this rejects one flawed model for another equally flawed model. Wittgenstein argues, 'It's [the inner] not a Something, but not a Nothing either! The conclusion was only that a Nothing would render the same service as a Something about which nothing could be said' (*PI*, §304).

In essence, we mistakenly demarcate the first-person use (a something; Cartesianism), from the third-person use (a nothing; behaviourism) of psychological verbs, which gives rise to conceptual confusions. In fact, the answer lies between these two extremes: 'The expression of a mental process is a *criterion* for that process; that is to say, it is part of the concept of a mental process ... that it should have a characteristic manifestation' (Kenny 2004: 49), or as Nagel (1986: 22) opines: 'The conditions of first-person and third-person ascription of an experience are inextricably bound together in a single public concept.'

## 5.3 Using the asymmetry principle to establish a category error in how educational predicates are used

### 5.3.1 A definition of a category and category errors

A *category* is a collection of objects, together with a collection of rules which govern these objects. A *category error* therefore, is the incoherent ascription of attributes to an object which cannot possibly have these attributes.

A rather famous example of a category error is Wittgenstein's discussion of sticks being neither awake nor asleep. Indeed, as Wittgenstein argues, to suggest that a stick is not asleep, is *not* to suggest that it is awake. This example seems simple, yet absurd. The error here is in assuming that one can ascribe the attribute of 'awake' or its negation 'asleep' to a stick. This error is a category error, because sticks do not belong to a category whose objects are governed by the rules 'awake' and 'asleep'. Therefore, the claim that a stick can be either 'asleep' or 'awake' is *incoherent* – not false – in light of this category error.

## 5.3.2 The first-person/third-person category error in education

Suppose now, these category notions are transposed to the realm of learning and, in particular, the sub-disciplines, mind, brain and education, brain-based learning and neuroeducation. It is the primary focus of this chapter to investigate the pitfalls of positing that the brain is the *location* of learning, thinking, understanding, etc.

Taking the brain as the locus of psychological phenomena induces a profound category error, on the basis of first-person/third-person asymmetry. The only thing which takes place within the brain is brain activity; not learning nor any other psychological predicate for that matter, for example knowing, perceiving, thinking, experiencing. This brain activity cannot be replaced by the predicate 'learn', as the process which takes place inside the brain.

However, like the stick being neither awake nor asleep, it also does not follow that the brain does *not* learn. Rather, it makes no sense, on the basis of this category error to house learning *or its negation* inside the brain. That is, learning is neither inside the brain, nor outside it; It is not a something (an inner guiding 'object' inside the brain), and not a nothing (behaviour only, with no inner activity) either.

### 5.3.2.1 The asymmetry principle for educational predicates

Recall, the notion of first-person/third-person asymmetry entails an understanding that there is an oppositional difference in the way we ascribe the attribute to the first-person (ourselves), in contrast with how we ascribe the same attribute to the third-person (someone else).

So, for example, in this instance when we consider the predicate 'learn', we ascribe this predicate differently to ourselves as to another. Indeed, in claiming '*I have learnt*', I can make such a claim about myself without requiring an appeal to some sort of criterion. That is to say, classrooms are not full of children opening their books in astonishment when they realized that they got 10 out of 10 in a set of problems they had no idea they had learnt. Such evidence is not required for the first-person to ascribe learning to himself. The predicate 'learn',[1] therefore, is ascribed to oneself *without criteria*.

To talk, however, of *someone else* learning, is ascribed *with criteria*; perhaps some form of observation about their behaviours or actions, in accordance with some set of predetermined criteria to outline success or failure, that is, 'agreement' with these criteria or otherwise.

The predicate 'learn', therefore, is ascribed to oneself *without* criteria, but to another *with* criteria. In this way it follows the asymmetry principle.

The first-person and the third-person are, however, inexorably linked (Bennett and Hacker 2003; Nagel 1986). Indeed, should someone display an inability to ascribe learning, for example, to *another* in their third-person ascriptions, it is reasonable to assume that their first-person ascriptions of learning *to themselves* may also be misguided. The first- and the third-person usages are entangled; never to be pulled apart in an attempt to understand one as separable from the other.

### 5.3.2.2 *The symmetry principle applied to brain activity*

On the other hand, the notion of first-person/third-person symmetry is the idea that the ability or attribute would be ascribed *in the same way* to the first person as to the third person. So, for example, the activity in a brain at any given time is governed by a first-person/third-person *symmetry*, since the observation of activity in the brain can be done for the first-person in the same way as for the third person (by observing neural images e.g. PET and fMRI scans), and consequently the ascription of brain activity to oneself is conducted *in the same way* as for another.

### 5.3.2.3 *Different categories*

Therefore, it is clear that psychological predicates – particularly the educational predicates 'learn', 'understand', 'think' – and brain activities are in different categories, on the basis of their first-/third-person symmetry/asymmetry distinctions; and any attempt to place psychological predicates inside the brain in place of (i.e. as if it were the same as) brain activity induces a category error.

---

[1]  The predicate 'learn' is taken as an example. All other psychological predicates work in the same way.

That is, the predicates 'learn', 'think', 'understand' cannot possibly be housed inside the brain since they do not have the appropriate construction, and are not governed by the same rules. Any brain-based educational theories, therefore, which are predicated on the notion that the brain is *where* psychological phenomena can be housed inside, are founded on this profound category error, failing to grasp the nature of the predicates involved.

## 5.4 Concluding remarks on mereology and asymmetry

Taking the mereological fallacy and the asymmetry category error together yields an outright rejection of the notion that the brain is either the agent or the locus of psychological phenomena. This is a damaging blow to the neuroscience movement in general, and to the collaborative work which takes place within education and neuroscience in particular.

Despite various efforts to salvage from the wreckage a modicum of conceptual clarity (most notably in Searle and Dennett's respective efforts in Bennett et al. 2007), it is this author's view that such attempts have led only to further confusions, and reinforcements of the same conceptual concerns already outlined.

Both Dennett and Searle, for example, put forward the idea that, although the brain may not learn, there are various 'sub-systems' of the brain which display learning-behaviour and that claiming that these sub-systems 'learn' is harmless (Bennett et al. 2007: 112).

Dennett attempts to obviate the problems of the mereological fallacy in particular with his argument, *The Intentional Stance*, in which he claims that the 'attenuated forms' of psychological terms *can* intelligibly be ascribed to the brain and to its sub-systems, and thus give rise to 'sort of definitions' and 'sort of explanations'. We are to think of a brain and the parts of the brain as a subject of the intentional stance, meaning it (the brain) and its parts can be regarded as rational entities capable of this attenuated form of intentionality. The brain and its parts, it is argued, has partial desires and 'pseudo-beliefs', it *sort of* decides and 'demi-learns'. And what might it look like for the brain or its parts to do any of these 'attenuated' things? 'Don't ask', says Dennett (1998: 362).

Furthermore, Searle argues that claiming that the brain is the locus of thinking and learning, for example, is no more conceptually dangerous than claiming that the stomach is the locus of digestion (Bennett et al. 2007: 112).

These attempted retorts to the mereological fallacy and the asymmetry category error have various responses, and it is largely unnecessary to cover old ground in these concluding remarks on the matter. What must be said however is that these attempted solutions seem to give rise only to further confusion.

The 'sub-system' argument is no more plausible than the 'brain as agent' argument, since if the mereological fallacy applies to the brain, presumably it must also apply to *parts* of the brain, which are themselves, by definition, also part of the human being.

Dennett's sub-system, intentional stance argument, which gives rise to 'sort of definitions' and 'sort of explanations' is hardly a solution at all, since it does not resolve any of our concerns in any meaningful sense, and Dennett himself seems quite happy to admit that he is not overly concerned with conceptual clarity, rather he is focused on grasping at whatever insights he can gain with regard to the brain. The abuse of language, therefore, Dennett argues, is simply something which might just be necessary in order to keep the investigation going.

And finally, Searle's argument that thinking can be seen in the brain in the same way as digestion can be seen in the stomach is confused in its very own particular way. It ignores asymmetry, and in fact adds to the confusion surrounding the principle. Searle is right that digestion can be seen in the stomach, and one might see that if the stomach was opened under anaesthetic. He is wrong to say that thinking, like digestion, can be seen in the brain. What does a thought look like? Do we understand this question? Is there any way to answer it? Furthermore, digestion is a physical process governed by symmetry of ascription, whereas thinking, as it has been shown, is governed by asymmetry of ascription. Searle, rather conveniently, overlooks this logical distinction. In doing so, he contributes to a further case of the asymmetry category error.

The conclusion, therefore, is clear: the brain is neither the agent nor the locus of psychological predicates. It is incoherent to posit either that the brain learns or that learning takes place inside the brain. Neuroscience has undoubtedly contributed to this conceptual confusion, and has done very little to obviate the subsequent complications since Bennett and Hacker (2003) outlined them.

Brain-based learning theories seem to hold – tacitly or otherwise – that the brain is either the agent or the locus of learning, thinking, etc. It should trouble us greatly that such profound conceptual blunders underpin much of what is under investigation in this collaboration.

# Neuroscience and Irreducible Uncertainty

## 6.1 Introduction

In this chapter, one of Wittgenstein's more counter-intuitive ideas will be invoked and applied, it is safe to say, for purposes other than what it was originally intended. The eventual aim is to apply these ideas to the language and use of psychological predicates and attributes (e.g. learning, understanding, thinking) and how they can be ascribed to human beings intelligibly and coherently.

This chapter seeks to show that these educational attributes and their ascriptions are governed by irreducible uncertainty,[1] while also adhering to the first-person/third-person asymmetry principle outlined earlier in this book. It will be shown that psychological phenomena such as believing, thinking and understanding cannot be *reduced* to brain activity, thus undermining the neuroscientific hypothesis that brain events give rise to, or are equivalent to psychological phenomena.

So, for example, the aforementioned claim from neurophilosophers like Searle that, 'with the development of fMRI and other imaging techniques, we are getting closer to being able to say *exactly where* in my brain the thoughts occur' (Bennett et al. 2007: 109, emphasis added) will be subjected to further philosophical scrutiny.

Moreover, let it be eminently clear that this section *does not seek* to outline that neuroscience (and in particular neuroscientific techniques such as neuroimaging, for example) tells us nothing about the brain or the activity which takes place within it. Such an assertion would be simply incorrect. Rather, the core remit is to demonstrate how Wittgenstein's PPF §315 can be applied to support the case that neuroscience *serves no purpose* in reducing the uncertainty which is a constitutive feature of (at least) some third-person ascriptions of psychological

---

[1] Also known as 'constitutive uncertainty'.

predicates – such as believing, learning, thinking and understanding. Thus, it will be argued that, whatever neuroscientific techniques and methods *do* show, it cannot possibly be learning, thinking, believing, understanding and the like.

Moreover, in keeping with Wittgenstein, it will be shown that 'the language we use to talk about the inner[2] will remain *irreducibly uncertain*, since there is no way to move to a position of certainty' (Purdy and Morrison 2009: 103).

## 6.2 Neuroscience, education and irreducible uncertainty

In Part II of *Philosophical Investigations*, known as *Philosophy of Psychology – A Fragment*, Proposition 315, Wittgenstein makes the following radical claim, which breaks from mainstream philosophical thinking about the nature of 'knowing' one's own mind, in contrast with the mind of another:

> I can know what someone else is thinking, not what I am thinking. It is correct to say 'I know what you are thinking', and wrong to say 'I know what I am thinking.' (A whole cloud of philosophy condensed into a drop of grammar). (*PI – Part II PPF*, §315)

Now, what might the implications for education be if such a claim is adopted to educational practices; and, in particular how might such a proposition impact on the potential collaboration between neuroscience and education? – a collaboration which is predicated on the belief that neuroscience can be used for educational purposes to reduce the uncertainty of concepts like learning, thinking and understanding, through neural imaging examinations of the brain.

Indeed, as this author has already demonstrated, there is considerable evidence to demonstrate that it is commonplace within neuroscientific and educational discourse to seek to *reduce* the uncertainty of psychological predicates such as 'learn', 'think', 'understand' and 'believe' by examining the brain.

The premise of such efforts is simple: it is believed that with advancements in neural imaging techniques that what was once inaccessible is now fast-becoming accessible. Moreover, where behaviour was once seen as a poor substitute for what goes on 'inside' the person, neuroscience is beginning – or so we are told

---

[2] Purdy and Morrison (2009) cite 'the inner' in their section. This citation induces no ambiguity, since the 'inner' in Wittgenstein's writings is treated as the hub of the psychological predicates which are of interest to this section. Rather than spend extended amounts of writing space defining the 'inner' according to Wittgenstein, let it be marked as 'read' that the arguments made in Purdy and Morrison (2009) are equally apt for the purposes of this section.

– to shed new light on the neural instigators of such behaviours. The uncertainty, therefore, surrounding behaviour is becoming *reducible*.

Curricula are now shaped by 'neuroeducational' thinking. Educational policy is now informed by the latest findings from within neuroscience. The brain, it seems, is taken to be the learning, thinking, understanding organ; the hub of all our educational endeavours – an intuitive, yet strangely misleading and dangerous assumption. Governments are giving increasing air-time to neuroscientific projects, to inform educational policy and so-called best-practice or evidence-based practice.

This section, by examining one of Wittgenstein's most unusual propositions, seeks to establish whether or not such centrality of neuroscience within educational discourse is founded. It will be argued that many of the neuroscientists cited previously fall prey to yet another fundamental error; namely, they seek to reduce the uncertainty of how we speak about psychological predicates by reducing (i.e., logically connecting) these predicates to brain states. That is, the core remit of this section is to demonstrate that learning, thinking and understanding, for example, *cannot* be described as states of the brain (or the mind, for that matter). Therefore, it will be contested that neural imaging – the staple of the neuroscientific method – *serves no purpose* in reducing the uncertainty of how we discuss psychological predicates, since what is shown on the neural scan (brain states) cannot be 'logically linked' (Purdy and Morrison 2009: 105) to the predicates in question.

The feeling of this section is captured in Wittgenstein's warning in *Zettel*, where he claims that this interchangeability of brain states and psychological states can lead to inherent problems. The author quotes at length, to avoid any misrepresentation:

> No supposition seems to me more natural than that *there is no process in the brain correlated with associating or with thinking*; so that it would be impossible to read off thought-processes from brain-processes. I mean this: if I talk or write there is, I assume, a system of impulses going out from my brain and correlated with my spoken or written thoughts. But why should the *system* continue further in the direction of the centre? Why should this order not proceed, so to speak, out of chaos? The case would be like the following – certain kinds of plants multiply by seed, so that a seed always produces a plant of the same kind as that from which it was produced – but *nothing* in the seed corresponds to the plant which comes from it; so that it is impossible to infer the properties or structure of the plant from those of the seed that comes out of it – this can only be done from the *history* of the seed. So an organism might come into being even out

of something quite amorphous, as it were causelessly; and there is no reason why this should not really hold for our thoughts, and hence for our talking and writing. (*Z*, §608, this author's italics)

In this proposition, several key points are contributing. Wittgenstein is not saying anything about the existence of such unusual plants which seemingly ignore the usual causal picture. He is, rather, creating a thought experiment which challenges us to think differently about the nature of causality, in particular about the nature of what we might call '*local*' or '*intrinsic*' causality. He is challenging our intuition about the way we are led to think about the 'inside' of an entity, on the basis of what it presents on the 'outside'. That is, when presented with different outputs, the temptation is to generate a model of explanation (which is generated in local causality) which gives rise to the belief that a difference in the (unseen) inner 'caused' the difference in the (seen) outer.

In relation to brain processes – which are worthy of special mention in this proposition, such is their importance – Wittgenstein is challenging the belief that brain processes cause thought processes. He warns that such a view of things neglects to examine the *non-local* aspect of the causal picture; namely, the '*history*' of the interactions and role which this history plays in the full causal nexus.

Wittgenstein's proposition calls out to the modern-day neuroscientist to warn him that the brain *may not* be where the causal story ends. If, as this author suspects, Wittgenstein is right, the inherent value of neuroscience – in particular, the value of the neuroscientific method – for education in providing a fuller *understanding* of the psychological processes in question is fundamentally limited. The purpose of this chapter is to examine whether these attempts to establish a parallelism between neurophysiological processes and psychological processes have any philosophical credibility. Wittgenstein seems to have concluded that any hope for such a parallelism ought to be distinguished:

It is thus perfectly possible that certain psychological phenomena *cannot* be investigated physiologically, because physiologically nothing corresponds to them. (*Z*, §609)

By examining also PPF §315, it will be shown over the next few sections that such warnings ought to be adhered to very closely.

There are three principles embedded into these discussions by Wittgenstein, which when taken together, offer insight into the profound implications of PPF §315 for neuroeducation in particular and brain-based learning theories in general, with reference to the overarching principle of constitutive/irreducible uncertainty, namely (1) first-person/third-person asymmetry (again); (2) a categorial distinction between 'know' and 'certainty' and (3) a subtle grammatical distinction between 'can' and 'cannot'.

## 1. *First-person/third-person asymmetry (again)*

It is clear in PPF §315 that Wittgenstein is making a sharp distinction between the first-person (the 'I' in 'I am thinking') and the third-person (the 'someone else').

On the basis of this distinction, what Wittgenstein seems to be saying here is that one ascribes 'thinking' (and other such psychological attributes, for that matter) to oneself in a manner which is fundamentally different from how one ascribes thinking to others.

Recall this sharp distinction is found in Wittgenstein's discussion of criteria: one ascribes thinking, for example, to oneself without reference to criteria, whereas one depends on criteria in order to make meaningful ascriptions to others. One point of note, important for Wittgenstein's discussion of criteria in particular, is the notion that these criteria, when appealed to, are in fact public constructs, which act as norms or maxims for the meaningful ascription of psychological attributes to other people.

Furthermore, as it has been argued previously, one's criterial third-person ascriptions are inexorably tied to one's non-criterial first-person avowals, insomuch that if one displays consistent inaccurate ascriptions of thinking, for example, to others in one's third-person ascriptions, then the credibility of one's first-person avowals may also be called into question. In this sense, the credibility of one's first-person statements also depends on the consistent displays of mastery of the criteria required to make one's third-person ascriptions.

## 2. *A categorial distinction between 'knowing' and 'certainty'*

It is clear from Wittgenstein's later philosophical work, in particular that found in *On Certainty*, that he makes a careful delineation between what it means to 'know' and to be 'certain'.

With typically careful consideration for the grammatical connotations of each of these terms, Wittgenstein makes the subtle argument that knowing is admissible in cases when to doubt makes sense, whereas, in contrast, certainty follows from the logical exclusion of doubt. In this sense, knowing and certainty are considered to be categorially distinct.

Moyal-Sharrock (2007: 80) outlines that, 'For Wittgenstein, it is *not* true that a mistake gets more and more improbable as we go from a hypothesis to a rule.

At some point a mistake has "ceased to be conceivable"[3]. A change of category, not of degree, has occurred'.

Similarly, Glock (1996: 77) outlines that Wittgenstein suggests that 'objective certainty … signifies the inconceivability of doubt or of one's being mistaken', going on to claim that such objective certainty 'belongs to a different category from knowledge'.[4]

For Stroll this sets Wittgenstein up as a form of foundationalist[5]:

> Wittgenstein's genius consisted in constructing an account of human knowledge whose foundations, whose supporting presuppositions, were in no ways like knowledge. Knowledge belongs to the language game, and certitude does not. The base and the mansion resting on it are completely different. That is what Wittgenstein means when he says that knowledge and certainty belong to different *categories*. (1994: 145)

Finally, Malcolm serves to further distinguish between Wittgenstein and Descartes, while highlighting the inherent similarities which exist in both of their respective understandings of knowledge and certainty:

> There are two points of agreement between Descartes' conception of 'metaphysical' certainty and Wittgenstein's conception of 'objective' certainty. Both conceptions have these two features: first, being certain of something, in either sense, entails the impossibility of being able to seriously entertain any doubt about it; second, this certainty entails the impossibility of one's being mistaken. Descartes' conception has the third feature that this certainty entails that one *knows* what one is certain of. But in Wittgenstein's conception this is rejected. (1986: 211)

Stroll (1994: 148) emphasizes this position, claiming that 'Wittgenstein, as a foundationalist, also asserts that nothing could be more certain than that which stands fast for us', going on to outline that, 'One cannot sensibly ask of that which is certain whether it is known (or not known) or true (or false); for what is meant by certitude is not susceptible to such ascriptions'.

This position is clear in PPF §315, with Wittgenstein's use of the word 'know'. He says, 'I can know what someone else is thinking', which on the face of it seems in line with his concept of knowing, since it is possible to doubt what someone

---

[3]   Cited from *On Certainty*, §54.
[4]   Cited from *On Certainty*, §§54–6, 193–4, 308 and *Last Writings on the Philosophy of Psychology, Volume 2:88e*.
[5]   See Stroll (1994: 147) for the philosophical definition of foundationalism.

else is thinking. However, Wittgenstein also claims that it is wrong to say 'I know what I am thinking.' Why? Quite simply because one's first-person ascriptions, being non-criterial avowals, are not open to doubt. Therefore, one cannot know one's own thinking; one is certain of it.

*3. A subtle grammatical distinction between 'can' and 'cannot'*
Over a series of propositions in Wittgenstein's Last Writings (I and II), he puts forward various arguments which seem contrary to what has been argued in the previous two subsections, most notably in (*LW I*, §§243, 951, 963, 964, 967; *LW II*, MS 169:33e).

For the sake of clarity, consider how Wittgenstein observes:

> 'The *uncertainty* as to whether another person is in pain' – is it based on the fact that he is he and I am I? (But ask yourself: 'Can he know it? He doesn't have any object for comparison.') No, *here* I'm deceived by a picture. The uncertainty is a matter of the particular case, and the concept vacillates from one case to another. But this is our game – we play it with an *elastic* tool. (*LW I*, §243)

This proposition seems to capture the feeling which is hidden also in the subtleties of PPF §315. It is clear that there does not exist a one-to-one correspondence between the first- and third-persons and the distinction between certainty and knowing. Indeed, in the third-person case in particular, the uncertainty, as Wittgenstein remarks above, is 'a matter of the particular case'. In essence, this means that third-person ascriptions are *context dependent*. For example, as Wittgenstein notes:

> There are cases where only a lunatic could take the expression of pain, for instance, as sham. (*LW II*, MS 169: 33e)

Such a view of things is consistent with Wittgenstein's use of *criteria* in third-person ascriptions of psychological attributes. The game – played with 'an elastic tool' – has the tendency to oscillate between the particular cases, where third-person ascriptions are complex and intricate. There are cases when doubting the displays of outward behaviour and doubting their sincerity would conflict with the consistency of other aspects of one's world view. In such cases, doubt is instinctively suspended, and one takes the displays of behaviour as certainties.

In other words, when Wittgenstein suggests in PPF §315 that 'I *can* know what someone else is thinking' he is saying that there are cases when knowing *is possible*, but that there are also cases when it is not possible, and others when one may be certain.

In the example of third-person pain ascriptions, Wittgenstein is saying that if one sees a person crippled with pain having had their arms and legs cut off, for example, one *is certain* that this person is in fact in pain, because to doubt that a dismembered person is in pain is not a part of our world-picture. This context-dependent nature of criteria is a feature of third-person ascriptions, which Wittgenstein says is played with an 'elastic tool'.

In the first-person case, however, the picture is entirely clearer: by saying 'I *cannot* know what I am thinking', Wittgenstein is clearly saying that it is *never* possible to know one's own thinking, since one cannot doubt it.

### 6.2.1  PPF §315 explained

It follows, therefore, that by taking first-person/third-person asymmetry, the categorial distinction between knowledge and certainty, and the grammatical difference between 'can' and 'cannot' together one arrives at the full picture of what Wittgenstein means in PPF §315. One can 'know' what others think, since, in some cases, the way in which one ascribes thinking to others (in the third-person sense) is open to doubt. One cannot 'know' one's own thinking since it cannot be subjected to doubt, in the sense that it is ascribed in the first-person ascriptions in a non-criterial manner. In first-person ascriptions, the situation is certain (i.e. indubitable) and non-criterial; in third-person ascriptions, the situation is, in at least some cases, open to doubt and invokes the use of criteria. Fusing together the asymmetry principle, the categorial distinction between knowledge and certainty, and the difference between 'can' and 'cannot', yields Wittgenstein's PPF §315.

Hacker (1986: 277) concludes that what Wittgenstein is claiming in PPF §315 is 'deeply counterintuitive and lacking any obvious rationale'. Why? Well, according to Hacker it is because 'one is viewing it from the wrong angle' (ibid). The error, therefore, according to Hacker (in keeping with Wittgenstein) is that we are prone to abusing the language of the first- and third-persons, as well as the grammar of epistemology (1986: 277–8). Knowledge claims require a greater deal of care, and the distinctions between the first- and third-persons must be negotiated, keeping in mind the asymmetry principle of ascription. The impact of such philosophical thinking on the educational landscape, however, is yet to be established. That is the focus of the next section.

## 6.2.2 The implications for brain-based approaches to education

What then, are the implications for neuroscience and its impact on education of PPF §315? In particular, what role, if any, can be played by the use of neural imaging in resolving the uncertainty of what someone else is thinking? The simple answer, unfortunately, is: none.

The profound impact of PPF §315 on discussions about psychological predicates is that the proposition transcends the boundaries of science, in particular, modern-day science about the brain. Wittgenstein's original proposition was inevitably aimed at discussions about the mind, rather than about the brain. Indeed, processes such as learning, thinking and understanding were considered, at Wittgenstein's time of writing, to be faculties of the mind. Modern-day science, however – in particular cognitive psychology and neuroscience – has demonstrated the tendency to transcribe such faculties onto the brain in crypto versions of mind–body dualism, which now manifest as modern-day versions of brain–body dualism (Bennett and Hacker 2003; Hacker 2012). This attempt then gives rise to examinations of the brain using neural imaging techniques – which have become the staple of the neuroscientific method – in the hope of reducing the uncertainty of the psychological predicates in question.

Wittgenstein, however, in his writings was pre-emptively suggesting such attempts are senseless, to the point that they induce a *category error*:

> It is tempting to try to demarcate the domain of ordinary psychological concepts by invoking the contrast between the mental and the physical and identifying the psychological with the mental, on the assumption that the mental and the physical constitute two exhaustive, mutually exclusive realms. (Bennett and Hacker 2003: 117)

However, Bennett and Hacker (2003) go on to conclude that 'this is mistaken, and the dichotomy is unhelpful' (ibid). The category of the psychological cannot be so simply reduced to a mental realm, whose apparent uncertainty could be reduced by a closer examination of the 'inside'. When Wittgenstein claimed that it is wrong to say 'I can know what I am thinking,' whereas it is right to say 'I can know what you are thinking' he was not claiming that such 'knowing' and 'not knowing' were a result of some uncertainty which could be reduced. A deficiency of 'internal' facts was not the source of Wittgenstein's qualm. I can 'know' what someone else is thinking means there are cases when I *cannot be certain* of what they are thinking; that is, such uncertainty is constitutive and irreducible when it exists. I cannot 'know' what I am thinking means I am certain of it. Therefore, what Wittgenstein meant was that the uncertainty of our ascriptions of psychological predicates to others (in our third-person ascriptions) is a constitutive feature of these ascriptions, in the cases when uncertainty exists; a feature which does not present itself in our first-person ascriptions of the same predicates to ourselves.

Searching the brain in order to reduce this uncertainty is incoherent. Delving inside the skull to look at synapses fusing and neurons firing does not reduce the uncertainty of the psychological; simply because the category of the psychological obeys the asymmetry principle, which, as a consequence, means that one's first-person ascriptions are *avowals* which are beyond doubt (i.e., certain), and one's third-person ascriptions are *descriptions* which are irreducibly uncertain, in the cases when such uncertainty exists.

Glock (1996: 177) observes that Wittgenstein's philosophical psychology, particularly his later writings, does not permit the transgression from the 'mentalist' (mind–body dualism) view to the materialist (brain–body dualism) view. Despite the fact that the materialist view seems more plausible than the mentalist view – by virtue of the fact that it induces a physical entity (the brain) in place of some ethereal realm (the mind) – Wittgenstein, nevertheless, dedicated some of his later writings to dispensing with anything more 'up-to-date'. The hope that the neuroscientific method (neural imaging) can reduce the uncertainty which has plagued the 'hidden' inner realm is pre-emptively halted before it can gather any pace. As Glock (1996: 177) concludes:

> it seems plausible that mental phenomena are inner causes of outward behaviour, and must hence be identical with neurophysiological phenomena, that is, brain-processes or -states. However, even if one grants this causal conception of the mind, it does not follow that psychological statements describe neurophysiological phenomena. If Wittgenstein is right, first-person present tense psychological utterances are by-and-large not descriptions of anything, let alone the brain. Less controversially, what little I know about my brain is based on fallible evidence, but that I have certain sensations, intentions, beliefs, etc., is neither subject to error, ignorance or doubt, nor based on evidence or observation of any kind.

There is clearly a category difference between the psychological and the neurophysiological, evident in the fact that the language used to talk about both realms is not interchangeable. Consider, for example, the following string of statements, used by Glock (1996: 178):

A1: I am in pain
A2: my C-fibres are firing
B: I can doubt whether my C-fibres are firing

Notice that the use of A2 in statement B leads to no apparent confusion. I can indeed doubt that my C-fibres are firing, since such an event is based on fallible evidence, which I may or may not be convinced by, of have knowledge of. However, Statement A1 cannot be readily substituted into statement B in place of A2, since it would then read:

B*: I can doubt that I am in pain,

which is clearly a nonsense. What might it mean to doubt that one has such-and-such a pain. Even in cases of severe hypochondria, the person who claims to be in pain exhibits no doubt about their pain, regardless of its lack of substance. Their actions demonstrate a lack of doubt. The problem of 'substitutability' gives rise to a difference of categories between psychological states (such as pain, learning and thinking) and brain states. Whatever is shown on the brain scan or neural image, therefore, cannot be logically connected to a psychological state.

To be clear, this does not amount to a denial that a properly functioning brain is a 'precondition for the possession of mental capacities' (Glock 1996: 178). Moreover, the lack of logical connection between psychological states and brain states does not preclude that there exists a *correlation* between some mental phenomena and neurophysiological states. However, Wittgenstein's philosophy gives rise to the notion that there is, as a matter of fact, 'a universal parallelism between the mental and the physical' (ibid).

Glock (1996) cites a series of propositions which this author cited previously from Wittgenstein's *Zettel* which seem to capture this feeling, namely §608–9 and §611. In these propositions, Wittgenstein concludes that the yearning for a causal picture between the psychological and the physiological should be abandoned.

As a result of these realizations, we arrive at the conclusion that 'even where neurophysiological phenomena are, as a matter of empirical fact, correlated with mental phenomena, they are neither necessary nor sufficient for the latter' (Glock 1996: 178). That is to say, even if brain state A has, in the past, correlated to psychological state A*, this in no way suggests that being in brain state A *always* means being in psychological state A*. One is in psychological state A* only when one is in that state.

For example, the brain state for being in pain does not mean that one is in pain. Rather, one is 'in pain' when one is in a state of pain, not when one is in 'brain-state pain'. Conversely, it is logically possible to be in a psychological state B, even if the neural image does not correlate. So, for example, one could be thinking that things are thus-and-so, even if the neural scan suggests a contradictory brain state.

It is therefore logically impossible to establish necessary or sufficient brain states for psychological states; and this logical impossibility results from the categorial distinction between psychological phenomena and neurophysiological phenomena. The conclusion is, therefore, 'There is no conceptual connection between neurophysiological mechanisms and mental phenomena' (Glock 1996: 179). Moreover, linking this to PPF §315, 'Neurophysiological concepts play no role in our explanation and application of mental terms: third-person uses of mental terms are based on behavioural criteria, first-person uses are not based on any criteria, let alone neurophysiological ones' (ibid). The uncertainty surrounding our third-person ascriptions of psychological predicates is, as a consequence, irreducible, through neuroscientific methods or otherwise.

It is important to note that *we cannot cross the categorial boundary between knowledge and certainty with the use of neuroscientific methods.* The uncertainty is *irreducible*. This 'type' of uncertainty is sometimes labelled in the philosophical literature as the 'constitutive indeterminacy' of mental concepts (ter Hark 1990: 147–52; Hacker 1993: 136–41), meaning that the uncertainty results not from a deficiency of information at the observer's disposal; rather, the uncertainty is unresolvable, since it is embedded in the nature of what is being observed or investigated. It is 'constitutive' since further examination – by whatever means – will not resolve the uncertain or indeterminate nature of mental phenomena. It is a feature of the philosophers' attempts to salvage a causal picture of how the inner and the outer are related to try to resolve this constitutive uncertainty. An unwarranted – yet understandable – commitment to the causal laws of the mental causing the behavioural gives rise to the misguided view that it is a mere deficiency of information about the observed entities 'inner realm' which means we ought to search further.

The techniques and methods of cognitive neuroscience and neurobiology have been born from this misconception. The mentalist, Cartesian-type statement 'If only I knew what was going on inside him' has been replaced by the materialist, neuroscience-type statement 'If only I knew what was going on in his brain', which leads to neural examinations and scans in an effort to reduce our uncertainty. But this uncertainty is unresolvable in the case where it exists. It is a feature of what is being measured. It is not, as Wittgenstein remarks, 'a shortcoming' (*RPP II*, §657).

Neural imaging, thus, brings us to an impasse insomuch that such techniques are used to resolve the uncertainty which is a constitutive feature of the psychological predicates in question. Therefore, any hope of reducing the uncertainty of the nature of psychological predicates of interest to education by looking at the brain is incoherent as well as redundant. Suter (1989: 152) concurs, warning us what the bounds of philosophy permit us to do:

> Philosophers go wrong ... when they try to reduce psychological concepts to something else or to identify psychological phenomena with brain events, overt behavior, disposition to behave, or with anything else. You cannot say truly '*That* is what F is', where 'F' is a psychological term. What philosophy should do, instead of trying to identify or explain the phenomena of mind, Malcolm observes, is to '*describe language*. It should remind us of what we say. It should bring to mind how we actually use the mental terms that confuse us philosophically'. When we do this, we see why behaviorism, identity theory, and Cartesianism dualism must be rejected. We also find that psychological concepts are family-resemblance concepts and there can be no illuminating theory of the mind. (Suter 1989: 152)

Suter, in line with the core arguments of this section, underpins the previous claim with justification found in the first-person/third-person asymmetry of how psychological predicates are ascribed.

> This asymmetry in the use of psychological and mental predicates ... we must take as one of the special marks of the mental. Physical predicates display no such asymmetry.
>
> Some people regret this asymmetry. They wish that the grounds for saying that others are sad were the same as our reason for saying that we are – namely, because we feel sad. After all, we say that Jones is six foot tall for the very same reason we say that we are. ... We have to accept the logic of our language, including the logic of our psychological language. (Suter 1989: 152–3)

It is a feature of our first-person ascriptions that they are non-criterial, non-observational and certain. Such ascriptions are *avowals*. Conversely, it is a feature of our third-person ascriptions that they are criterial, observational in relation to others' behaviours and, in many cases, *only knowable*; that is, they can be doubted. Such third-person ascriptions are *descriptions*. The uncertainty of our third-person ascriptions in such cases is constitutive of such ascriptions and thus irreducible; an examination of the brain serves only to trade one form of uncertainty for another. On the other hand, the certainty – that is, the unintelligibility of doubt – of our own first-person ascriptions is already steadfast, in need of no further improvement, by neuroscientific methods, or otherwise.

Part Three

# The Philosophy of the Inner and the Outer: Neuroscience, Cartesianism and Mind-Brain Identity Theory

# Inner and Outer: The Epistemology of the Mind

## 7.1  Introduction to Part 3

In this part of the book, the author will develop some of the philosophical ideas from Part 2, with a particular focus on the epistemology of the mind. There will also be a focus on offering philosophical critiques of Cartesianism and mind-brain identity theory, which, as this author will show, form major parts of the neuroscience story.

## 7.2  Chapter introduction

Within the bounds of this chapter of the book, the author will examine several prevalent themes of Wittgenstein's writings on the so-called inner/outer picture, which highlights the connections between the inner and the outer according to Wittgenstein.

The author will examine the surrounding literature on matters such as epistemic privacy, direct and indirect access, the epistemology of the inner/outer relation, and mental and behavioural phenomena, with the hope of establishing a cogent argument against Cartesianism, and all of its guises, including modern-day theories such as materialism and reductionism. This chapter, therefore, serves as a timely reminder of Wittgenstein's continued arguments *against* the Cartesian philosophy of the time, which was fundamentally committed to dichotomizing the inner and the outer into two autonomous, separately analysable realms.

## 7.3 The inner/outer picture

In the entry entitled 'inner/outer' (or, *Innen/Außen* in German script) (Glock 1996: 174–9), the core strands of the inner/outer picture which are prevalent in Wittgenstein's work are discussed. There is an examination of what the 'inner' *is*, as well as what the relation is between the inner and the outer, and the mental and physical; all of which characterize the so-called inner/outer picture. Glock (1996) also examines what temptations we are often drawn towards when we speak about this relation between the inner and the outer, and why these temptations are strangely alluring.

Consider, first, the most tempting understanding of the inner/outer relation. As this author has already outlined, it is intuitive (perhaps?) to suppose that the inner stands 'behind' the outer, causing behaviours and *both* volitional and non-volitional actions alike. The inner in this way, surely, is the source of the outer, a private, hidden realm to which only its owner has a true – some say 'privileged' – access. The inner, therefore, under this view of things, is an entirely mental realm, independent of the outer world to which it gives rise (or causes).

Furthermore, it is commonplace for the philosopher to identify the inner with the concept of the 'self' (or the 'I'); the aspect of the person which is considered to be subjective and private, to which only the owner has some privileged access (Hacker 2010: 266).

This author has already offered the beginnings of a critique of this (apparently) intuitive view of what the inner 'is' and how it relates to the outer. Glock (1996: 174) suggests that the temptation is to believe that the inner/outer picture is generated within dualism, where the 'mental' and the 'physical' are captured by the 'inner' and the 'outer', respectively, Glock (1996: 174–5) continues:

> We find it natural to distinguish between the *physical* world containing matter, energy and tangible objects, including human bodies, which is *public*, and the human mind, a *private* world *hidden behind* our behaviour. And we think that each individual has a *privileged access* to his own mind, while our access to the minds of others is indirect, based on observations of their behaviour, and at best uncertain. Wittgenstein regards this as a 'picture' which is embedded in our language (emphasis added).

Rather strangely, Cartesianism seems to have embedded itself rather deeply into how the inner is spoken of in everyday terms also. The intuitive nature of dualism – what is a Cartesian trait – is something which we feel compelled to defend, tacitly or otherwise, despite the rejection of Cartesianism as a philosophical doctrine:

> It is by no means merely in Cartesian theories that talk of inner objects and process abounds; our everyday psychological language is filled with such phrasings, too. We talk about calculating in the head, for example, insist that we clearly see a situation before us when discussing a past event, and worry that while one of our friends looks perfectly happy on the outside, he is in fact terribly unhappy within. (Bax 2012: 35)

Consequently the inner/outer relation is a 'picture' which emanates from the use/misuse of language. As Bax (2012: 35) outlines, the problems induced by the flawed Cartesian inner/outer picture disappear only when philosophy detaches itself from the literal application of the inner and treats it as anything other than a metaphor.

> According to Wittgenstein, the problem with Cartesianism rather is that it takes our talk of inner objects and processes too seriously – and that it thereby, on a different level, by far does not take them seriously enough. As several of his remarks make clear, portraying psychological phenomena such as thoughts and feelings as literally inner entities actually fails to capture what we take to be their essence.

The notion of the inner as a collection of objects, processes and hidden entities is cast aside, adding philosophical credibility to the infused relation between the inner and the outer, rather than the dualist detached one put forward within Cartesianism. This rejection of dualism – a theme of Wittgenstein's philosophy – will be a fundamental cornerstone of this chapter of the book.

### 7.3.1 The relation of the inner/outer picture to the asymmetry principle

This application of care in how mental and psychological predicates such as 'learn', 'think' and 'understand' are used, is a re-occurrence of the asymmetry principle outlined earlier in this book. As Glock (1996: 175) suggests, the crux of the inner/outer illusion is found 'in the fact that we apply mental predicates to others, but not to ourselves, on the basis of behavioural criteria, something "external"'. This claim is the essence of the first-person/third-person asymmetry principle. Therefore, the inner/outer picture confusions unfold and become resolvable, only when the asymmetry principle is adopted; and, conversely, all philosophical problems generated within the inner/outer picture result from an ignorance or a rejection of this principle.

Within philosophy, as long as one restricts oneself to speaking about the inner/outer relation in a manner which is governed by the asymmetry principle, no conceptual blunders are likely to appear in how this relation is spoken of; however, as Glock (1996: 175) acknowledges, 'outside philosophy the distinction between the mental and the physical does not coincide with this dichotomy of inner and outer', citing, as an example, that 'we regard toothache as *physical* pain, to be contrasted with mental suffering'.

It is indeed tempting to regard the inner as a private, hidden realm, ethereal in nature, accessible only to its owner. Moreover, it is tempting to think that this realm, and the faculties it gives rise to, are the root causes for behaviours. The inner, surely, is the 'cause' of the outer; insomuch that the connection between these two realms is a local relation, with the inner (mind/brain?) supervening on the outer (body; both volitional and non-volitional acts/behaviours)? In fact, these are precisely the philosophical positions which Wittgenstein contests, and which this author will now show to be incoherent, however tempting they may be to accept.

The importance of the inner/outer picture in philosophy comes to light, when one considers the various strands of philosophical thinking which it informs. As Glock (1996: 175) argues, 'The inner/outer picture informs not just Cartesian dualism, but the mainstream of modern philosophy, including rationalism, empiricism and Kantianism.' The same author continues to outline that other philosophical doctrines such as idealism, phenomenalism, behaviourism and materialism also place significant interest in their respective considerations of what the true inner/outer relation is. One of these doctrines will be of particular interest for the bounds of this book – namely, materialism – when this author argues in Chapter 8 that neuroscience and consequently neuroeducation and brain-based learning are modern-day versions of crypto-Cartesianism, falling under the bracket of materialism, insomuch that properties that were once posited of the mind (under Cartesian dualism) are now posited of the brain (under neuroscience).

It is clear, therefore, that the inner/outer relation has dominated many strands of philosophical thinking, particularly those strands in which the relation between the mental and the physical is of interest. Subsequently, it is often accepted without question that the inner represents the mental world, and the outer represents the physical world, and these two autonomous, independent realms are connected causally. It is this view of things which requires closer examination. As Glock (1996: 175) outlines, citing Wittgenstein in the process,[1] 'The mental is neither a fiction, nor hidden behind the outer. It infuses our behaviour and is expressed in it.' This claim is at the centre of what Wittgenstein holds true of the inner/outer relation. Notice that this claim seems to exhaust all the options, and yet arrives at no clear conclusion. If the mental is not a fiction, then it must be real; that is, it must have some discernible qualities. However, whatever the 'mental' or the 'inner' *is*, it is not hidden; that is, it is not ethereal, nor is it a private realm hidden on the inside, *behind* the outer, as it were.

In fact, as this author has outlined in previous sections, it requires the outer for its very definition. The inner is intertwined with the outer, and the behavioural representations which are apparent in how one acts.

The use of the word 'infuses' in this instance seems rather apt. Indeed, in grasping that the inner and the outer are 'infused', one embraces the notion that neither can be conceptualized intelligibly without the other; but also that these two realms cannot be pulled apart to stand at a distance from one another. This *infusion* of the outer and the inner is quite literally an *extraction* of information from both realms, applied to the other. That is, the true inner/outer picture is one in which the inner is unintelligible in the absence of the outer, since in the absence of the outer, no such infusion – that is, no *extraction* – would be possible.

The infusion of the inner and the outer into one inseparable entity, undoubtedly serves as one of the cornerstones of Wittgenstein's considered assault on Cartesian philosophy. Where Cartesianism posits a dualist approach to the inner and the outer, the mind and the body, the ethereal and the real, Wittgenstein contests that such philosophizing leads only to confusion. The inner is not some hidden sanctum in which a private play takes place. More importantly, the inner does not stand behind the outer in a dichotomy of realms which, when pulled apart, stand together only via some 'spooky' metaphysical causal relation. Wittgenstein's philosophy gives rise to a sophisticated escape from the mantra of Cartesianism embedded in our everyday lexicon. Bax (2012: 43) concludes that only when the Cartesian dogma is dispensed with, can philosophy move forward with a greater degree of credibility; and the lynchpin to this progression is the adoption of Wittgenstein's philosophy:

> Instead of seeing the mind and body as diametrically opposed or only contingently related, Wittgenstein takes them to be intrinsically connected, thereby breaking

---

[1] Citations are given by Glock (1996), taken from '*The language of Sense Data and Private Experience – Notes taken by R. Rhees of Wittgenstein's Lectures, 1936*', 10–11, 134–5; *Philosophical Investigations* §357, *Philosophical Investigations, Part II*, 178, 222–3; *Last Writings on the Philosophy of Psychology Part II*, 24–8, 81–95.

the myth of the inner as a separate and closed-off realm. Freed from this preconception, a person's doings and sayings do not have to be set aside as purely external clues, and a sentence like 'She is over the moon' can consequently be taken to genuinely describe the state that another human being is in.

Such infusion and extraction can take place only between two compatible realms. In this view of things the entire make-up of the inner is *extracted* from the outer. Inner, mental faculties are *infused* with their outer representations. The inner is only 'inner' insomuch as it takes the language which is developed in the outer world, and applies it *inside*. Why is this even necessary? This author contends – in keeping with Wittgenstein (and indeed, with Quantum pioneer, Niels Bohr[2]) – that the inner *depends* on the outer for the language used to describe it, since there is no other language available. However, in using 'outer' language to describe and talk about the inner realm, one must recognize that it is not possible to talk about the inner in a way which makes objective *sense*, without reference to its *relation* to the outer world, from which this language is borrowed. In applying language from the outer to the inner, one entangles these realms together – or *infuses* them – to the extent that the inner, whatever it might be, is unintelligible in its own right. This inner/outer 'system', however, will never yield unconditional objectivity; rather, it will yield as much objectivity as is possible to extract from a system which does not want to exude total objectivity.

Notice that this argument is not equivalent to suggesting that every inner action has an outer parallel or vice versa. Rather, all that this argument puts forward is that the *language* of the inner is borrowed from the outer, but that this act of 'borrowing' means that the inner can only be spoken about *in relation* to the outer, from where the language is taken. The consequence of this *relational* – a keyword – system is that all 'inner' concepts are taken in relation to the 'outer'. Inner concepts are close cousins of outer concepts; that is, they bear a *family resemblance*. That is, a system where the 'inner' and the 'outer' are pulled apart to stand independently of each other is one where neither realm makes any sense.

The temptation might be to suggest that, if no language for the inner exists which permits the inner to be spoken of independently of the outer, then a new 'inner language' ought to be concocted. But this inner language, if fundamentally distinct from the outer language, would be a private language, which Wittgenstein railroads against as conceptually incoherent in his famous 'private language argument', the details of which will be examined later in the book.

---

[2]  Bohr's ideas in relation to the philosophy of quantum mechanics will be discussed in Part 4 of this book.

For the moment, however, it is sufficient to note that the use of the terminology of 'inner' does not mean 'private' or 'hidden from view'; since such inner faculties are constructed on the basis of presuppositions of outer world precepts. For example, one cannot calculate in the head (in one's 'inner realm'), if the concept of 'calculating' has no meaning in the outer world. In such an instance, what takes place 'in the head', whatever it might be, is inexorably infused with the outer concept of calculation. In the absence of the outer, the inner ceases to make sense.

Does this mean, therefore, that inner faculties are reducible entirely to outer behaviour? Glock (1996: 175) acknowledges that 'Wittgenstein's attack on the inner/outer dichotomy is often accused of reducing the inner to the outer, and thereby ignoring the most important aspects of human existence'. In essence, this attack on Wittgenstein's stance was taken to be the argument that he was a 'behaviourist in disguise'. Indeed, in general, the view that the inner and the outer are intertwined in this way is often confused for some version of behaviourism, be it logical, metaphysical or methodological.

In any case, it suffices to say that this take on the inner/outer relation – that is, the view that the inner and the outer are 'infused' or 'entangled' realms, intelligible only when considered as related – does not amount to behaviourism 'in disguise'. In fact, 'Wittgenstein in turn accuses the inner/outer conception of mistakenly assimilating the mental to the physical', in that it,

> construes the relationship between mental phenomena and mental terms 'on the model of' material 'object and designation', and thereby turns the mind into a *realm* of mental entities, states, processes and events which are just like their physical counterparts, only hidden and more ethereal. (Glock 1996: 175)

The conclusion is, therefore, that the 'inner' is not simply a collection of mental 'faculties' which are constructed in the outside world and taken inside the mind (or brain for that matter).

Now, this may seem at odds with what this author has already purported of the inner/outer relation. Indeed, only a few paragraphs back this author contended that the inner is only 'inner' insomuch as it takes the language of the outer 'inside', to describe the inner in terms and concepts 'borrowed' from the outer world.

However, to extend Wittgenstein's position on the basis of this claim, the argument is that the inner is *not* merely a collection of objects or processes which take place in the mind or brain that can be 'carried around' and 'called upon' whenever required. For such a view of things would suggest that the inner realm is a collection of archived (mental?) materials, which are the entire essence of us as human beings, the very thing that defines us as what we are; whereas, in fact, this take on the inner is 'like Platonism … fuelled by the Augustinian picture of language, which suggests that all words stand for objects, and all sentences describe something' (Glock 1996: 175).

This is rejected, and this rejection is perfectly in keeping with previous claims about the inner/outer picture. In essence, the connection between the inner and the outer is grammatical, not epistemic nor causal. The error arises when, like Descartes, and indeed like Locke, one attempts to connect the inner and the outer in a local causal relation, where the inner is epistemically private (a key term, returned to in the next section), and ontologically hidden.

The conceptualization of the inner/outer picture is central to understanding further aspects of the Wittgensteinian assault on Cartesianism, which is the focus of the remainder of this chapter, in considering concepts like epistemic privacy, privileged access and mental indeterminacy.

## 7.4  Wittgenstein on epistemic privacy and privileged access

Another indictment of the flawed Cartesian inner/outer picture is found in Glock's discussion of Wittgenstein's dismissiveness of so-called epistemic privacy, an aspect of Cartesianism which was subsequently reinforced by Locke in his conception of introspection as inner perception (Hacker 2010: 245). Glock (1996: 176) captures what Wittgenstein means by this concept: 'Wittgenstein turns on its head the idea of epistemic privacy, according to which only I can know that I am in pain,[3] while others can at best surmise it.'

Epistemic privacy is also sometimes known as *privileged access*. It is noteworthy that in both Bennett and Hacker (2003: 89) and Hacker (2010: 246), there is a discussion of the notion of 'concealing' one's mental workings, where such acts of concealing give rise to the misunderstanding that the inner is hidden from view, private and accessible only to its owner. This is where the metaphor of privileged access begins; in the assumption that *only* we can know our own minds. Such access, therefore – it is assumed – is private and privileged by virtue of the fact that it is only open to the person whose mind it is.

---

[3]  Being 'in pain' is Glock's example cited here, and it is one often used by Wittgenstein himself. The argument, however, extends to all psychological predicates such as, in particular – for the bounds of this book – learning, thinking and understanding.

This misconceived view of the inner/outer picture is of interest to this book because it engrains and embeds Cartesian thinking further into discussions about all psychological predicates, such as learning, thinking and understanding. What distinguishes Wittgenstein from his philosophical opponents is his commitment to overthrowing Cartesianism in all its guises and forms. This commitment is eminently evident in his refutation of epistemic privacy.

Glock (1996) labels this misunderstanding as 'epistemic privacy', since it captures the apparent 'private' nature of what we know about our own mind. It must be acknowledged that it is particularly tempting to assume that one's mental workings *are* epistemically private, insomuch that they *seem* to be hidden from the views of the onlooker, who only gains access to the workings of another's mind if that person permits them a glance in the form of behavioural manifestations. It seems, as Bennett and Hacker (2003: 89) note, that only when the concealed is revealed, the apparently hidden inner workings of the mind are made clear to the onlooker.

To be clear, 'To be able to say how things are with one (subjectively speaking) is not to have *access* to anything, it is to be able to *give expression* to something' (Hacker 2010: 246). Rather unfortunately, due in the main to Descartes and Locke, this temptation is oft-fallen for, in the main, one imagines, due to its intuitive nature.

Nevertheless, the temptation of this alluring view of things ought to be ignored, since it is predicated on the concept of the Cartesian mind, which Hacker (2010: 247) claims is 'an aberration'. This claim is developed further, and outlined in detail in Glock (1996: 176–7), summarized below:

1. *First- and Third-Person Knowing*
There are two equally problematic issues left behind as a residue in philosophy which are symptomatic of Cartesian thinking, and are part of Wittgenstein's charge against epistemic privacy: first, there is the notion that only *I* (the first-person) can *know* my own mind, a notion which has been discussed in part in Chapter 6, Section 1.2 of this book, but requires further investigation now inside the context of Cartesianism; and, secondly, the converse notion that others (the third-person) only have, at the very best, a second-rate guess-type information about the minds of others, which manifests in philosophical terms as not *knowing* what goes on 'inside of him'. This is, rather unfortunately, what one subscribes to which one embraces epistemic privacy; namely, that third-persons ascriptions of mental faculties and psychological phenomena are, by definition,

founded on guess-work (Bax 2012: 40–1). It is sensible to explore each of these aspects of this feature of epistemic privacy in turn, briefly.

First, consider the Cartesian notion which holds that only the first-person can truly *know* his own mind. As Glock (1996: 176) observes, there is, of course, an innocuous use of the word 'know' in such instances when one wants to state with clarity that one really does *know* something. But these are not the cases to which Wittgenstein, for example, takes exception, nor is it the focus of this author's critique. Rather, the focus here ought to be one what *knowing* entails, and what it means for someone to 'know' anything, in particular about themselves.

As it has been shown in Chapter 6, Section 1.2 of this book, there are categorial distinctions between knowing and certainty which are clearly a feature of Wittgenstein's charge against Cartesian thinking. Knowing, it has been argued, goes hand-in-hand with doubting. So, in what way can one be said to *know* one's own mind, as the Cartesian would suggest? Indeed, as Bax (2012: 40) notes, this is the defining feature of Cartesian dogma, that 'first person certainty [is] to be a matter of infallible knowledge, owing to a person's having direct access to his or her thoughts and feelings by means of introspection.'

But then, the corollary question is posed: Can one be *mistaken* or be in any *doubt* about their own thoughts, feelings and ruminations? That is, if one says that one can *know* one's own thoughts, then one must concede that it is possible for one to have *doubts* about one's thoughts.

The same thing is true of attributes and predicates of interest to education, such as learning and understanding, for example, which fall prey to the same linguistic conundrum. Glock (1996: 176) concludes, therefore, 'Because there is no such thing as misperceiving one's one pain, or being mistaken about it, to say that I know that I am in pain is either a nonsense, or an emphatic assertion that I *am* in pain.' So too with educational attributes, then; to say 'I know that have learnt how to add' is either a nonsense or an emphatic assertion or one's having learnt to add. The Cartesian dogma, however, which holds that knowing one's own mind is evidence of epistemic privacy, is rendered incoherent.

In the converse direction, the Cartesian notion that the third-person has only an underprivileged glimpse at the inner workings of another's mind is also rendered a nonsense, when one considers the nature of the predicates in question. As Glock (1996: 176) notes, carrying on his pain example borrowed from Wittgenstein, 'in the ordinary sense of "know" others can, and often do, know that I am in pain'.

What this tells us is, precisely, that if Wittgenstein is correct – which is this author's contention – then the entire Cartesian programme centred on epistemic privacy has things *precisely the wrong way around*! Wittgenstein captures this neatly in a series of propositions in *PI*, §§246–50, where he challenges the entire essence of epistemic privacy governed by the Cartesian confusions surrounding the first- and third-persons:

> In what sense are my sensations *private*? – Well, only I can know whether I am in pain; another person can only surmise it. – In one way this is false, and in another nonsense. If we are using the word 'know' as it is normally used (and how else are we to use it?), then other people very often know if I am in pain. – Yes, but all the same, not with the certainty with which I know it myself! – It can't be said of me at all (except perhaps as a joke) that I *know* I'm in pain. What is it supposed to mean – expect perhaps that I *am* in pain? … *This much is true: it makes sense to say about other people that they doubt whether I am in pain; but not to say it about myself* [this author's italics]. (*PI*, §246, original emphasis)

Wittgenstein, thus, concludes:

> The sentence 'Sensations are private' is comparable to 'One plays patience by oneself' (*PI*, §248)

Similarly in Wittgenstein's *Last Writings on the Philosophy of Psychology, Volume I*, we see where the error arises in thinking of first-person private 'knowledge' of psychological phenomena.

> The opposite of my uncertainty as to what is going on inside him is not *his* certainty. For I *can* be sure of someone else's feelings, but that doesn't make them mine. (*LW I*, §963)
>
> 'I can only guess at someone else's feelings' – does that really make sense when you see him badly wounded, for instance, and in dreadful pain? (*LW I*, §964)

The entire essence of what it means to 'know' and what it means to be 'private' – in any intelligible sense – is undermined by Wittgenstein as a nonsense, in the context of applying these terms to the inner. As Bax (2012: 44) observes, 'On Wittgenstein's non-observational, non-descriptive account of first person utterances … the possibility of doubt and error does not crop up in the first place and no such manoeuvers are required in order to preserve first person certainty.'

## 2. *Direct and Indirect Access*
In a similar manner to the complexities surrounding the epistemology of first- and third-person 'knowing', there is also the conundrum which manifests as the distinction between direct access – often said to be available to the first-person

only – and indirect access – suggested, on the contrary to be available to the third-person.

The complications become clearer in this instance, however, when one ponders what precisely 'direct' and 'indirect' mean. Indeed, by creating a sharp distinction between first-person direct access, and third-person indirect access to the thoughts, sensations and ruminations, etc., the Cartesian becomes embroiled in a war of words, which seems to point towards the idea that the third-person access is of a lesser quality than that of the first-person. This often leads to statements of the kind, "If only I knew what was going on *inside of him*, I would know how he really feels!" which rather unfortunately are born from the intuition that one's access to one's own thoughts is more privileged than one's access to others.

Rather worryingly, and perhaps most tellingly for the bounds of this book, the neurosciences seem to fall prey to the same conceptual error, inasmuch as their methods are centred around *reducing* some uncertainty, and in trying to make what *appears to be* indirect access, more direct. Indeed, by examining the brain, there is a tacit (or otherwise) assumption that the indirect access to the workings of the third-person can be made more like the direct access to the workings of the first-person. In this model, observable behaviour is taken to be a crude marker of what hides behind the veil.

Rather than distinguishing between direct access to one's own psychological phenomena and the indirect access to others', one is better served in distinguishing between the first- and third-persons in the asymmetrical manner which has been discussed previously. One's first-person access is non-criterial, and is avowed. It is not direct, nor is it privileged; it is categorially different. In the third-person sense, however, access to others' psychological states is not 'indirect', as opposed to being criterial, where psychological phenomena are ascribe on the basis of observable evidence of the behaviour on show.

It begs the question, then: Why the tendency to talk of first-person ascriptions as 'direct' and third-person ascriptions as 'indirect'?

The answer, as Wittgenstein makes clear, is the problem of claiming that the 'source' of our behaviours and the 'cause' of our psychological phenomena are hidden, taking the concept of 'understanding' as his example:

> Now we try to get a hold of the mental process of understanding which seems to be hidden behind those coarser, and therefore more readily visible, concomitant phenomena. But it doesn't work; or, more correctly, it does not get as far as a real attempt. For even supposing I had found something that happened in all those cases of understanding, why should *that* be the understanding? Indeed,

> how can the process of understanding have been hidden, given that I said 'Now I understand' because I *did* understand? And if I say it is hidden – then how do I know what I have to look for? I am in a muddle. (*PI*, §153)

Although Wittgenstein chooses 'understanding' as the predicate to examine in this instance, his analysis is equally apt for other such predicates and sensations, such as learning and thinking, or, for that matter, experiencing pain.

The 'coarser ... more readily visible, concomitant phenomena' to which Wittgenstein is referring are the behavioural 'outward' manifestations which are in plain view. In this proposition, it is clear that Wittgenstein is suggesting that we go in search of something 'deeper', hidden behind these more readily visible phenomena, to *explain* the phenomena which present themselves to us as onlookers. In this case, the behavioural manifestations of 'understanding' appear to be poor substitutes for the 'inner' source from which such understanding emanates. As a passive observer, with only an apparent 'indirect' access to the apparently invisible 'source' of such behaviour, one is left with a feeling of indirect access, in a belief that 'If only I could see into his mind' or 'If only I knew what was going on in his brain', then I could 'know' – that is, I could *really know* – what 'it' is that he understands. Wittgenstein, however, discards such seemingly intuitive beliefs to be 'a muddle'.

### 3. *Mental and Behavioural Phenomena*

Glock (1996) observes a third aspect of Wittgenstein's rejection of epistemic privacy which is found in respect to a discussion which this author has already examined in previous sections. For the sake of clarity, it shall be highlighted again in this new context. It is found in the categorial contrast between mental and behavioural phenomena. As Glock notes, with reference again to the example of pain and pain-manifestations:

> Nevertheless, one might hold, I cannot see the pain itself, only the behaviour which expresses it. But this is like saying that I cannot see sounds or hear colours. It indicates only a categorial distinction between mental and behavioural terms, not that statements involving the former are always inferred from those involving the latter. (Glock 1996: 176)

In a similar way, one can draw conclusions about concepts that might interest us in educational discussions. Indeed, one might hold, that I cannot see learning, or thinking, only the behaviour which expresses learning, or thinking. But this is to fall prey to a common misconception about the contrast between behaviours and mental or psychological phenomena; and Glock notes this absurdity with an equally absurd reference to 'hearing colours' and 'seeing sounds'.

To be clear, the absurdity of the notions of 'hearing colours' or 'seeing sounds' does not result because, as a matter of fact, we do not hear colours or see sounds. Rather, the absurdity is in ascribing the *possibility* of 'hearing' or 'seeing' to the concepts 'colours' or 'sounds', respectively. In a similar manner, one comes to realize that it is not that one cannot 'see' the mental version of pain. Rather, searching for it *makes no sense*.

The idea that pain is something inner, something hidden, or something to which only its owner has a direct, privileged access inside the mind or brain is the consequence of muddled reasoning. Such reasoning confuses the categories of mental and behavioural. One 'sees' pain when one observes someone who experiences pain, and makes their pain manifest in their behaviours. The face of this person need not be transparent to 'see' the inner workings of the mind while the person is in pain. The claim that pain can only be 'seen' or observed by examining the inner workings of the person whose pain it is, is as absurd as claiming that one can (or cannot) hear colours or see sounds.

Wittgenstein also outlines a similar rejection of this aspect of epistemic privacy making reference once again to the ability to 'see' or 'sense' someone else's psychological state of mind:

> But surely he could *see* them just as you and I do. – But the word 'sense' still isn't unobjectionable. – What do I perceive via sensation? In addition to the so-called sadness of his facial features, do I also notice his sad state of mind? Or do I *deduce* it from his face? Do I say: 'His features and his behaviour were sad, so he too was probably sad'? (*LW I,* §767)

And finally,

> 'His pains are hidden from me' would be like saying 'These sounds are hidden from my eyes'. (*LW I,* §885)

An acknowledgement of the categorial distinction between the mental and the behavioural – which has been discussed previously in relation to first-person/third-person asymmetry, and irreducible uncertainty – leads to a realization that behaviour is not simply a poor substitute for mental faculties, which seem to be at play behind the face, as it were. When one sees someone else express pain behaviour, one does not think, without good reason, that this pain behaviour is a poor substitute for what is going on 'inside' him.

Similarly, in the context of education, when the teacher sees the pupil sunk deep in thought, he does not pause for a moment and contemplate whether the pupil is thinking or not. Even in the case where the teacher is unsure of the semantic content of such a thought, the 'hidden' nature of the pupil's thought would easily be lifted, upon the teacher asking 'What are you thinking about?' When the pupil claims to have learnt their times tables, and demonstrates behaviour which backs up these claims of 'learning' the teacher does not assume – again, without good reason – that they need to find out what is going on behind the eyes when this learning takes place. It would, in fact, be entirely wrong to assume, as Glock observes, that statements about the mental can be made only on the basis of inference of behavioural observations. Statements about the mental are, simply, statements about the mental. They are not inferred from statements about the behavioural; indeed, this view of things is to confuse a difference in *category* between the mental and the behavioural with a difference in *kind*. This is simply not the case. This is why Wittgenstein observes that the statement 'His pains are hidden from me' (i.e. '*His* pains are something mental, something hidden') is as absurd as the statement 'These sounds are hidden from my eyes'. The problem is not that his pains are hidden – that is, mental – rather, it makes no sense to look 'inside' for the mental concept of pain, when the concept of pain is to be found only in the behavioural manifestations which are in plain view.

## 4. *Epistemically Concealed, not Private*

The fourth and final confusion surrounding epistemic privacy to which Wittgenstein dedicates some of his philosophical writings on psychology is the notion of the fallibility of observations of others' behaviours. The fact that the evidence at hand *can* be misinterpreted by the observer or, for that matter, can be concealed by the person whose behaviour it is, gives the impression that the inner is a hidden, private realm and that the epistemology of this realm is open only to the owner of it.

Bennett and Hacker (2003: 89) label this aspect of Wittgenstein's discussion on epistemic privacy as the 'concealing metaphor', and other aspects of the discussion are also found in this author's discussions on the subtle grammatical distinction between 'can' and 'cannot' hidden inside Wittgenstein's PPF §315, outlined in Chapter 6. For the sake of clarity once more, it seems sensible to link these previously made arguments to the material in this section. Glock outlines this last facet of epistemic privacy and Wittgenstein's contestation of it, thus:

> It is tempting to protest that the mind is hidden in that there is always the possibility that others are lying or pretending. This shows that our third-person judgements are fallible. It does not establish the sceptical conclusion that, in a particular case, we are or could always be mistaken. Lying, deceit and pretence

are parasitic on sincere avowals of the inner: pretending to be in pain is not behaviour without mental accompaniment. ... Nor is pretence possible in all cases. (Glock 1996: 176–7)

Wittgenstein nicely captures this notion with a moment of realization hidden inside a rather innocent and innocuous looking proposition: 'If everything goes normally, no one thinks of the inner event which accompanies speech' (*LW I*, §120). The conclusion is, therefore, that one searches for the inner 'accompaniment' only in the cases where something seems to be hidden. In the cases where behaviour is overt, demonstrable and easily interpreted, one does not seek further clarification of a person's psychological 'state' by wondering what is going on 'inside' him.

The fact that someone else can conceal their true feelings from others leads to the assumption that their feelings are private; truly private in the epistemic sense. But this is not the case. Indeed, for such psychological phenomenon to be epistemically private, it must be beyond access to another, and accessible only to its owner. However, the philosophical problem disappears when one acknowledges that there are cases where the mental and psychological phenomena that seem interesting to us *can* be epistemically hidden, but in such cases they are not hidden because they are private by nature, rather they are *concealed*. The difference between a potentially concealed inner realm and a private inner realm is the source of the confusion.

## 7.5 The inner is NOT epistemically private

On the basis of these four aspects of Wittgenstein's refutation of epistemic privacy, it becomes clear that the doctrine of a private, hidden inner realm is philosophically problematic. Over the course of the last few sections, the author has outlined the inherent complexities which are embedded in many philosophical discussions surrounding the nature of the inner, and its relation to the outer. These complexities, it seems, are a residue left behind within mainstream philosophy from the time when Cartesianism was rife. Indeed, the notion of an epistemically private inner realm to which only the owner had direct access is a major aspect of Cartesian thinking surrounding the mind.

In relation to this book, dispensing with the flawed concept of an epistemically private inner realm is fundamental to this author's attack on Cartesianism, and its mutated and modern-day version which is hidden in the philosophy which underpins neuroscience. The conceptual links between the philosophy which underpins neuroscience and Cartesianism will be investigated in greater detail in Chapter 8, with the view of showing that Wittgenstein's defeat of Cartesianism is equally valid as a critique of neuroscience and its educational derivations today.

In the meantime, for the purposes of education, more specifically, if the notion of an epistemically private inner realm is cast aside, educators and educationalists alike can begin to move forward within educational discourse without the temptation to ponder what is 'going on inside' the heads of their pupils. If nothing is hidden, nothing should be searched for.

Moreover, if the inner is not private in any meaningful sense, then the value of sciences like cognitive neuroscience to education is fundamentally restricted. Indeed, if one is to follow a Wittgensteinian view of the inner, and – perhaps more importantly – a Wittgensteinian view of the *relation* between the inner and the outer, then one comes to realize that scanning the brain to find out what is going on behind the eyes while a child is learning, thinking, and so on, is of no importance to what one is seeking to understand.[4]

Learning, thinking and understanding, for example, are not epistemically private processes to which only the first-person owner has privileged, direct access. These processes are not hidden in the inner realm, whatever this might mean. Searching for the 'essence' of such processes in the inner, therefore – whether in the mind (Cartesianism, mentalism) or brain (neuroscience, materialism) – is incoherent, problematic and troublesome. With Cartesianism overthrown, and the staple of Cartesian thinking – namely, the private inner realm – shown to be incoherent, time would be better invested in a philosophy which bears none of the hallmarks of Cartesian thinking; in particular in the areas of psychology, education and science, which all remain – at least in part – committed to Cartesian dogma.

---

[4]  Which is not to suggest that neuroscience is valueless *per se*, rather it does not bring educationalists to some moment of clarity which would otherwise be unattainable.

# Inner and Outer: The Challenges of Crypto-Cartesianism, Materialism and Reductionism

## 8.1  Introduction

The *relation* between the inner and the outer has been the subject of discussion thus far. On the basis of these detailed discussions, two major arguments require further elaboration and discussion.

The first is the notion that neuroscience is a modern-day version of Cartesianism. This idea has been most notably developed by Peter Hacker, over the course of various works, most notably in two keynote lectures given in 2012 and 2014, and in various other works by the same author, in particular, Hacker (2010).

Second, there is a requirement to discuss Wittgenstein's pre-emptive defeat of materialism, which seeks to transcribe attributes once ascribed to the mind (in Cartesian philosophy) to the brain (neuroscience), thus resolving the inherent complexities of Cartesianism. This author will examine whether the reduction identification of the mind to/with the brain is plausible and coherent, and will therefore question whether the materialist account based on the brain is more acceptable than the conceptually flawed Cartesian account of mind. Needless to say, these two arguments are intrinsically linked.

## 8.2  Hacker on neuroscience as crypto-Cartesianism

The challenge of Cartesianism is as alive as ever. Despite extensive efforts, most notably by Wittgenstein, to overthrow the Cartesian doctrine, the core precepts of Cartesian thinking seem to find their way into our seemingly intuitive view of the world, our science and our philosophy. For the purposes of this book, and in order to draw timely connections between this author's discussions of the inner/outer relation and previous discussions about the brain, it seems sensible and pertinent to outline that neuroscience offers a classic example of a mutated form of crypto-Cartesianism, refreshed and renewed to face the modern-day challenges of anti-Cartesianism.

In a keynote lecture given in 2012 at St Anne's College, Oxford, Dr Peter Hacker delivered his talk entitled '*Are persons brains? The challenge of crypto-Cartesianism*'. In a further keynote lecture in 2014, Hacker made contributions to the debate entitled '*What can the brain teach us about the mind?*' In personal correspondence between this author and Hacker, the transcripts were made available. Some of the details of these transcripts will be used now as rationale and motivation for the ensuing discussion on neuroscience as a modern-day version of Cartesianism.

With Wittgenstein's defeat of Cartesianism established (cf. Chapter 7), the commitment of neuroscience to Cartesian principles should be cause for concern. Indeed, despite attempts to quash Cartesian links, neuroscience remains committed, at least in part, to a mutated form of brain–body dualism which seems to have replaced the defeated mind–body dualism posited by Descartes himself. In Hacker's typically candid style, he seems to capture the essence of this transposition from mind to brain, and argues why we ought to be *more* anti-Cartesian in what we posit about the human being and his behaviours (Hacker 2012: 4).

The Cartesian paradigm, Hacker contests, amounts to several troubling assumptions. Many of these assumptions have been discussed by this author already, in particular in the discussions of the inner/outer relation.

Hacker (2012: 3) outlines that the prevailing belief of Cartesianism is that, 'The mind is an immaterial substance that interacts with the brain at the pineal gland, at which point it controls, by means of acts of will, the flow of animal spirits to the muscles, causing them to expand or contract.' So, the mind supervenes on the brain and thus 'causes' bodily movement and behaviour. The human being is, in essence, controlled by the central hub which is the mind. The overt causal nature of Cartesianism is what troubled Wittgenstein himself. Why? Well, mainly because it led to the conclusion that, as Hacker (2012: 3) observes, 'Mental attributes are essentially privately owned and epistemically private.' This meant that, according to Cartesian thinking, 'The body a person has is causally related to the mind or soul that a person is.' Consequently Cartesian philosophy contends that:

> the mental is an attribute of the mind. It is the mind that is conscious or unconscious, that senses and perceives, that thinks and feels, that forms intentions, makes decisions, and that performs acts of free will that cause the body to move. The connection between the inner and the outer is causal, non-logical. (Hacker 2012: 3)

There is no doubt that this view of the mind–body dichotomy is tempting. Indeed, as this author has already outlined, this view of the inner/outer picture seems to be embedded in an intuitive view of mental, psychological and behavioural phenomena.

However, in keeping with Wittgenstein, this author has dispensed with the notion that the inner supervenes on the outer; and similarly that the mind supervenes on the body. The inner and the outer are infused, not separate. However, it seems that this dualistic approach is so deeply enshrined in philosophical doctrine that residues of it remain today. Hacker himself concedes that this view of things 'articulates a sophisticated form of fiction that grips the imagination of mankind', claiming 'It has cast a long shadow over European thought for the last four centuries' (Hacker 2012: 3). As Wittgenstein famously remarked to his student John Wisdom, 'What is troubling us is the tendency to believe that the mind is like a little man within' (Weinpaul 1958: 70).

Why is it then, that despite the constant and incessant rebuttal of these profoundly mistaken Cartesian ideas, 'the misconceptions to which it gave birth persist' (Hacker 2012: 6)? Why is it that these principles are so profoundly intuitive, despite their lack of sense, and why is 'mankind' drawn towards this philosophically flawed model? It seems that these concepts yield a model of supposed explanation which is more difficult to distance oneself from, than it is to accept.

Human beings yearn for explanations, and it is a part of our fabric to be drawn towards anything which looks vaguely like one. Perhaps more interestingly, at the basis of this yearning, lies a commitment to a causal picture which seems natural and intuitive. There is no doubt that Cartesianism is precisely that. It is a doctrine which appeals to our natural tendencies, a dualism which churns out first-person privacy and protects the apparent uncertainty of third-person ascriptions of psychological phenomena. The inner, as this author has argued, is posited to be hidden and private in this explanation of things, and the outer is at best a poor substitute for the first-person authority of self-ascriptions. The temptation is to be drawn towards a model which *seems* to protect all of the natural intuitions that are embedded in what way we view ourselves. Surely our mind controls our actions, and surely our thoughts take place inside this mystical realm inside the head. In the case where the mysticism of the mind – an apparently ethereal realm – is too much to grasp, doesn't it seem 'natural' to transpose our intuitions about this mind–body dualism into brain–body dualism in order to protect the intuitions embedded in our view of the world? We, as human beings, surely know our own make-up. Despite extensive efforts to recalibrate this discussion, we seem naturally drawn towards this dualist, causal picture.

## 8.2.1 The Cartesian mind, The Cognitive Neuroscientific mind and Aristotle's psuchē

Hacker (2010: 241 243, 255) gives a glimpse into the important attempts at conceptualizing the mind at three very different stages of the story, summarized in the table below, and at the same time offers insights into the glaring Cartesian nature of neuroscience:

| The Cartesian Mind | The Neuroscience Mind | Aristotle's psuchē[1] (non-Cartesian mind) |
|---|---|---|
| A substance | An organ (the brain) | Not a substance |
| Immaterial | Material | Neither identical nor distinct from the person/self/body |
| Identical with the person/self | Identical with the person | |
| Distinct from the body the person has | Distinct from the body the person has | Not a part of the human being |
| Part of the human being | Part of the human being | Informs the living organism but is not 'embodied' within it, that is, has no location |
| The subject of mental and psychological predicates (i.e. the mind thinks, etc.) | The subject of mental and psychological predicates (i.e. the brain thinks, etc.) | |
| An agent | An agent | The possession of a mind is the possession of faculties and powers of intellect and will |
| In two-way causal interaction with the body | In two-way causal interaction with the body | |
| The subject has a privileged access | The subject has privileged access to the qualitative content of the 'brain experiences' known as 'qualia' | Does not stand in a causal relation to the body |
| Concepts are privately owned and epistemically private | 'Qualia' are privately owned and epistemically private | Not an agent |
| Contents are indubitable to the owner | | Not the subject of psychological nor mental predicates |
| (p. 241) | (p. 243) | Not essentially private (p. 255) |

---

[1] The concept of psuchē in Aristotle's writings is an early precursor to the concept of mind as a collection of faculties of intellect and will. The psuchē, therefore, was not regarded as a possession (i.e. not a 'thing'); rather it was an idiomatic way of articulating the human being's rational and intentional faculties of intellect and will.

Neuroscience, it seems, unintentionally or otherwise, is the saviour of Cartesianism. It offers a way to salvage from the wreckage a mutated, crypto-version of dualist philosophy which remains fully committed to the dogma of the past.

Rather unusually, however, neuroscience was never intended to be a modern-day version of Cartesianism:

> What is most distinctive of the current intellectual and scientific scene is that despite its official materialism, and its much vaunted anti-Cartesian stance, it is, conceptually speaking, Cartesian through and through. Much of contemporary cognitive neuroscience repudiates the idea of the Cartesian mind, conceived as an immaterial spiritual substance, and avers that the mind just *is* the brain. It then proceeds to ascribe the attributes characteristic of the Cartesian mind to the brain. It is noteworthy that it retains intact the Cartesian conception of the structural relationship between the mind and the body, transferring it without alteration to the relationship between the brain and the body. (Hacker 2012: 3–4)

Be under no illusions, therefore, that Cartesianism is alive. It is, in fact, more dangerous given it is now hidden under a new name, and this name is oblivious to its Cartesian undertones. It seems that all neuroscience has served to do – philosophically and conceptually speaking – is to replace 'mind' with 'brain'. The ethereal 'mind' has been set to the side to be replaced by the material 'brain'. This is what this author has labelled 'the challenge of materialism'. Hacker (2012: 4) warns us that everything else has been left 'intact', claiming 'this is to succumb to the worst of the Cartesian misconceptions. For the mere replacement of the mind by the brain is altogether superficial. One needs to be much *more* anti-Cartesian. The question marks have to be put *much* deeper down.'

These arguments offer a significant connection between this author's work on neuroscience and the points made about the inner/outer relation thus far. It is clear that there are deep-seated confusions within the philosophical foundations of neuroscience, not only in the mereological fallacy and the first-person/third-person category mistake outlined earlier, but also insomuch as the philosophy which underpins neuroscience is Cartesian in nature.

It has been shown that Cartesianism attempts to pull apart the inner and the outer into two autonomous separable realms, connected only via a non-logical, local causal relation. Neuroscience falls prey to a similar conceptual mistake. The brain is no more a suitable candidate for the faculties once posited of the mind than the mind was itself.

To be sure, the blunder is at least more plausible than Descartes's dualist mistake. The brain is, in some ways, a more amenable entity to deal with; it can be examined, and its tangible nature makes it more natural to deal with. However, setting the mystical nature of what Descartes called 'mind' aside, and upon closer philosophical scrutiny, the brain is no more suitable. Over the next few sections, it will be clear why.

## 8.2.2  Replacing the Cartesian Mind with the Cognitive Neuroscience Mind/Brain Identification

There is no doubt that the brain is a *more intuitive* candidate for psychological attributes to be ascribed to than the mind is. The mind, insomuch as it is anything, is not tangible nor is it possible to investigate the mind in any meaningful sense. The brain, however, sidesteps these potential pitfalls, and it seems to offer a more justifiable account of the inner/outer dichotomy than mind–body dualism offers, for example. Let it be clear, however, that this was never the intention. Neuroscience is, as Bennett and Hacker (2003) and Hacker (2012) remark, overtly and deliberately anti-Cartesian. The dualism which is posited in Cartesian philosophy is not supported in neuroscientific discourse; not in any deliberate sense anyway. However, the residue of Cartesianism is stamped on neuroscience in what it posits of the brain in place of the Cartesian concept of mind.

The intuitive nature of the brain as a more sensible candidate than the mind for the ascription of psychological attributes ought not to lead us to believe that it *is* correct to assign these attributes to the brain. When this author concedes that the brain is more intuitive, let it be clear that what is meant is that it is more *excusable* why such erroneous ascriptions might be made, and the inherent conceptual problems are significantly greater to unravel. However, the problems remain.

Hacker (2012: 5) also outlines that the continued commitment to Cartesian/ Lockean philosophy which has 'dominated our thinking for the last three and a half centuries' has made this movement between the two forms of dualism more tempting. It is this author's contestation that this is precisely why neuroscience has become so successful at permeating mainstream thinking; namely, it infests on our seemingly natural intuitions about causation and the inner/outer picture. This author has already discussed the complexities with the inner/outer picture, and has gone to great lengths to show that, despite our intuition, the concept of a dualist, inner/outer dichotomy of two autonomous realms is flawed. Some initial discussion about causation has also already been put forward. Consider further what Hacker (1993: 135) says about causation:

> The dominant philosophical account of causation conceives of the causal relation as non-logical (external), inductive (hence requiring the possibility of independent identification of the relata), and nomic (instantiating a general law). The mental and its behavioural expression, however, do not unquestionably fit this account.

The relation between the inner and the outer, as well as the mental and the behavioural, does not fit the standard philosophical account of causation. Some effort will be spent over the remainder of the book to establish that this relation between the inner and the outer is a logical, non-local relation, where the relata cannot be independently identified thus invoking the concept of a relational attributes conceptual model.

Meanwhile, consider how Hacker (2012; 2014) offers various arguments which are central to his claim that neuroscience in general – and by this author's extensions, the educational derivations of neuroscience – are mutated forms of crypto-Cartesianism. Five main arguments will be considered from these papers, most pertinent to this discussion.

1. *Mind-brain identity theory;*
This is the notion that the mind and the brain are, in fact, conceptually equivalent. That is, the concepts of mind are (logically) equivalent to the concepts of the brain. This appeals to one's intuitions, since it gives scope to circumvent the apparently ethereal nature of the mind and focus more on the properties of the tangible brain.

But the mind cannot be logically (or otherwise) identical to the brain. The mind is not a thing, whereas the brain can be experienced with the senses. The brain has a spatial location, whereas the mind does not. It would appear the identity theorist is in logical difficulties: 'As Leibniz notes, if $a$ has a property that $b$ lacks, then $a$ is not identical to $b$. Now if $a$ is a physical event and $b$ is an emotion, $a$ will have the property of occurring somewhere, unlike $b$' (Suter 1989: 78).

Furthermore, the very nature of the question 'What is the mind?' seems, as Hacker (2010: 248) notes, to be 'a pernicious question', since it gives rise to an understanding of the nature of mind which is misconstrued. Such confusion is only eradicated when one considers, as Hacker (2010: 248–56) suggests, how the noun 'mind' is used in language. Then one will come to realize that the nature of 'mind' is simply a grammatical idiom used to describe faculties of intellect and will.

The mind-brain identity theorist, however, overlooks this intricacy and delves rather deeply into a quagmire of confusion, positing that the mind and the brain are conceptually identical or linguistically interchangeable. But Hacker's (2010: 249) grammatical analysis should warn us that such identification is not sensible. Indeed, in discussing memory, for example, one might idiomatically say that John has a poor memory by claiming that he is 'absent-minded' (Hacker 2010: 249). The problem of grammatical substitutability here gives a clear indication that mind and brain are not identical. Indeed, one would not be able to suggest that John is 'absent-brained'. Likewise with the concepts of *memory, thought, opinion,* and *intention* (Hacker 2010: 249), it makes perfect sense to invoke the idiomatic concept of 'mind' to describe such faculties, but one cannot likewise appeal to the concept of 'brain' in the same idiomatic manner. The noun 'mind' and the noun 'brain' are simply used differently; and such different use points towards different concepts.

Thus, it follows that brain activity (the physical event) cannot be logically identified with, for example, the emotion 'joy' (*Z*, §§486–7), on the grounds that the brain activity takes place inside the brain, whereas the emotion 'joy', is not located anywhere, other than, perhaps, the location where one experiences the joy; for example, at the theatre or restaurant. For this reason, Wittgenstein opines: '"I feel great joy" – Where? – that sounds like nonsense.' (*Z*, §486). One cannot ascribe 'joy' a location, in a manner which is akin to ascribing brain activity to the location of the brain.

Thus, since the activities of the brain (i.e. brain activity) cannot be identifiable with the faculties which constitute the idiomatic use of the concept of 'mind', it follows that the mind and the brain are neither logically equivalent nor conceptually identical.

## 2. *The brain is seen as a type of computer;*

Various analogies are developed along these lines in modern-day neuroscience, but they have existed in the brain sciences for many years. As Hacker (2012: 5) notes, there was a time when brains were compared to a telephone switch board, which was simply the most advanced technology of the time. 'We have always been prone to try to render ourselves intelligible to ourselves in terms of our latest technology.' However, Hacker warns us against taking this analogy with technology too far, claiming, 'This modern analogy is no deeper than the older one. Taking it seriously leads to such intellectual aberrations as computational psychology, which masquerades as cognitive psychology' (Hacker 2012: 5–6).

Such analogies tend to lead only to further conceptual confusion, and such confusion is even more dangerous when it happens to underpin empirical research: 'If conceptual howlers underpin empirical research, there is no hope to reach empirical understanding of what puzzles us' (Hacker 2012: 12).

There are many neurophilosophers; however, none more so than Searle for example, who argue for the development of such analogies, in order to elucidate previously unknown facts about the human brain. A particularly good example of such arguments is found in Searle (1990), an article for *Scientific American*, entitled '*Is the Brain's Mind a Computer Program?*', arguing that the *brain's mind* (whatever that is) is in fact *more sophisticated* than a computer, on the grounds that it can interpret information, and attach meaning to it, whereas a computer can only manipulate syntax. These conclusions are confused in their own very particular way, none more so than claiming that the view that *the brain*, let alone the human being, which has a mind, and that it is this mind which does what human beings would normally be held to do.

Moreover, Searle (1990) cites what he regards as four self-evident 'Axioms', which he takes as the underlying assumptions of his thesis. They are: (1) Computer programs are formal (syntactic); (2) Human minds have mental contents (semantics); (3) Syntax by itself is neither constitutive nor sufficient for semantics; (4) Brains cause minds. (Searle 1990: 27–9). From these, Searle draws four conclusions: (1) Programs are neither constitutive nor sufficient for minds; (2) Any other system capable of causing minds would have to have causal powers (at least) equivalent to those of brains; (3) Any artefact that produced mental phenomena, any artificial brain, would have to be able to duplicate the specific causal powers of brains, and it could not do that be just running formal a program; and therefore, (4) The way that human brains actually produce mental phenomena cannot be solely by virtue of running a computer program (Searle 1990: 27–9).

These conclusions should appear more than a little troubling, and a far distance away from any kind of serious consideration with regard to what the mind is, and what role the brain plays (if any) in shaping mental phenomena. More worrying, however, are Searle's so-called Axioms, which for a formal logician are taken as self-evident truths in need of no justification. Rather conveniently, Searle 'proves' his conclusions by taking a carefully selected collection of Axioms, some of which[2] are *in themselves* statements requiring proof and are *ipso facto* non-Axioms, which ultimately means that his 'Conclusions' are simply *ignoratio elenchi*.

Indeed, central to Searle's main conclusion that the brain's mind is more sophisticated than a computer program, are two fundamental assumptions, captured over all four Axioms: first, the notion that the mind is a 'thing' which *can* be caused; and, secondly, the brain's mind is capable of doing things that would normally be ascribed to a human being, that is, attaching semantic content to the syntax it receives. These are claims very much in need of proof. They require evidence. *They are not Axioms.* Searle, rather unfortunately, provides no such evidence. He goes very much in the opposite direction of what Hacker (2010) suggests is the 'nature of mind' and castigates – rather unsuccessfully it could be said – the notion that the mind is simply an idiom used to describe the collection of human faculties of intellect and will. If the mind can be caused, then Searle is appealing to the word 'mind' in a very *real* way, not the idiomatic way in which Hacker (2010) suggests it should be.

This author concludes further that Searle may also be guilty of *argumentum ad populum*, in that he appeals to forms of argumentation which are enshrined in popular culture and pop-/pseudoscience, (i.e. the brain causes the mind, or that the mind is something which 'just follows' from the brain/brain activity) as a basis for his seemingly far-reaching conclusions. This 'Axiom' is tempting,

---

[2] Axioms 1 and 3 are reasonable, although the use of the word 'Axiom' is still questionable. Computers are indeed formal in that they can only manage syntax (combinations of 1s and 0s). Moreover, syntax is not constitutive nor sufficient for semantics, and this claim may well be connected to a famous logical proof known as *Gödel's Incompleteness Theorems*, the details of which are omitted. Axioms 2 and 4, however, are far from self-evident, and there is no empirical evidence to support them. Axiom 2: 'Minds have mental contents (semantics)' is a questionable conceptual claim, since it is itself predicated on a model of mind which is undefined in Searle's paper, and which seems to tacitly suggest that the mind is a treasure trove of artefacts which are stored somewhere on the 'inside'. This use of the word 'mind' is far from the use of the noun as it appears in ordinary language. Quite naturally we might say, idiomatically, that 'My mind is full of other things today,' as an excuse to our wife for forgetting to collect the children from school, or that 'I'll keep it in mind' to our friend who invites us to a surprise party next April. But in neither case, nothing is 'inside' the mind, and the mind cannot be said to meaningfully have any content. When Searle says minds have contents, he is suggesting that they *contain* something, that something is *stored*. This is not an Axiom. Perhaps a better word for what occurs when we invoke the concept of mind is *retention*, not *storage*. Whereas storage demands a spatial location, retention does not. Therefore, we might say that a human being uses his mind to retain information, and to use it and reflect upon it, rather than saying, as Searle does, that minds have contents. What he means by Axiom 4: 'Brains cause minds' is even more unclear, and seems to fall prey to the Cartesian and Lockean concept of mind which is dispensed with by Wittgenstein and Hacker with great success. This notion is, in fact, the materialist equivalent of Cartesianism, which is one of the most glaring examples of Hacker's (2010, 2012, 2014) argument that the mind–body dualism of old has been replaced by the brain–body dualism of today. Searle simply goes a stage further, and invokes mind as the intermediary between brain and body, and suggests that the brain has a mind, which it causally supervenes upon.

but dangerous, and entirely anti-axiomatic in nature. Searle, it seems, may well have been swept along by the cognitive neuroscience tide, and these conclusions result from assumptions which are far from self-evident or trivial.

In any case, the overarching hope, it seems, of the appeal to such computer analogies is that if such models are developed in line with what is better understood – in this case the workings of a computer – then a fuller understanding of the brain will also follow. This author does not subscribe to such philosophical methods.

### 3. *The excitement around neural scanning;*

As Hacker (2012: 6) opines, 'the unwarranted excitement and credulity about fMRI scanning and its apparent results fosters the thought that we can actually *see* the brain thinking, perceiving, feeling emotions and making decisions.' This discussion has already been covered in detail in Chapter 4, Section 1.2.5 of this book, outlining precisely what neural scans *do* show, rather than what some neuroscientists purport they show. Nothing further need be said at this point.

### 4. *The identification of the person with his brain;*

Hacker (2012: 6) notes further that it is also common to hear of thought experiments being developed and cited to offer support to the view that the human being simply *is* his brain. Such thought experiments are, in fact, rather famous: varying degrees of sophisticated science fiction, involving discussions about brain transplants and brains being put in a vat.

The point of these thought experiments is to show that in the extreme case where the brain is detached from the body in which it would normally be housed, the person whose brain it was goes with it. That is, by removing the brain (and storing it in some form of 'vat', where the brain is connected to a computer screen and is kept 'alive' with sophisticated scientific techniques not yet available to us) it is conceivable that the person whose brain it is, is simply transferred from his body into the vat.

The standard '"brain" in a vat'[3] experiment in particular gives rise to a view that the human being simply is his brain; that his most compelling features which make him, him are kept inside his brain, and as such when this brain is reconnected to some sophisticated device which serves the same purpose to

---

[3]   This thought experiment has been invoked in various forms by many philosophers through time. A famous parallel of the more modern 'brain in a vat' argument is Plato's 'Allegory of the Cave' argument, which was a philosophical experiment used to examine the details of knowledge and reality, and how our internal beliefs and ideals construct our view of the world.

the brain as the body would in normal circumstances, it seems that the person remains, for all intents and purposes, intact.

What these thought experiments give rise to – other than a piece of science fiction – is the incorrect and incoherent conclusion that 'You are your brain' (Hacker 2012: 7). However, as Hacker (2012: 7) warns: 'Rather than be swept along by the tide, we should pause, and go back to the beginning – the beginning of conceptual enlightenment, which is clarification of what exactly is meant by terms we use and abuse with such enthusiasm.'

The most significant passage which seems to be quoted in reference to this topic is due to Wittgenstein, who claimed:

> Only of a human being and what resembles (behaves like) a living human being can one say: it has sensations; it sees, is blind; hears, is deaf; is conscious or unconscious. (*PI*, §281)

If Wittgenstein is correct in claiming that only to a human being or what 'resembles' a human being can one ascribe the characteristics and predicates of a human being, then one must wonder: In what way does the brain resemble the *entire* human being? Or, as Hacker (2012) questions: Can one identify oneself with one's brain?

There are three basic arguments which point towards Wittgenstein being correct in his assertions in *PI* §281. First, there is the physical argument, which states that since the brain is something which the person 'has', it cannot possibly be equivalent to the person. That is, how would it be intelligible to suggest that one can be the same as something which one has?

Secondly, there is the grammatical argument, and the lack of linguistic substitutability which points to the lack of identification between brain and person. For example, it makes sense to say, and it often is said, that 'John has a brain', but it is a curious statement to claim that 'John has a person'. Rather, one might say 'John *is* a person,' but not that he 'has' a person. Moreover, part of us referring to John as a person, seems predicated on the notion that John has a brain. In this way, there seems to be some grammatical differences between the nouns 'person' and 'brain' and their ordinary usage in everyday life. Therefore, in extending further the previous point, it is incorrect to identify an object which 'is' to another object which it (i.e. the other object) 'has'. In other words, if P is a property which is part of the criteria for the object A being identified as being 'A-like', then it is incorrect to identify the *property* P (e.g. having a brain) as being identical with the *object* A (e.g. being a person).

Thirdly, there is the psychological argument, which stipulates that the brain in a vat cannot possibly act like a human being. Such an assertion, tacitly embedded into the brain in a vat experiment, *is* inconceivable, despite the contestations of the supporters of such science fiction. Central to this argument is the notion that if the human brain is placed into a vat which can keep the brain fully functional, then the cyborg-type entity which remains would be the same as the person who had previously 'housed' the brain, now housed inside the vat. To begin with, such an assertion would *not* identify the brain with the person who previously owned the brain, which is oft-cited as the conclusion of this thought experiment. Rather, it might show that the cyborg-type brain in a vat (i.e. the combined brain + vat entity) is capable of behaving like the person who previously owned the brain. For such a claim to be credible, one would have to depart from the thought experiment, and prove that such an extension is scientifically possible. That is, at this point, we have moved beyond the point of 'thought experiment' and towards the realm of real science, which requires evidence and scientific rigour. In the absence of such evidence or rigour, such claims must remain as mere hypotheses at the very most.

More tellingly, perhaps, it is important to insert a modicum of sensibility into the discussions at this stage. Such sensibility it captured in the finer details of *PI* §281. The cyborg-type brain in a vat is simply not capable of the vast array of complex behaviours that one would normally ascribe to a human being. The brain itself cannot exercise linguistic mastery, and it is not the correct entity to which one can meaningfully ascribe the host of volitional faculties of intellect, will and reflection that one might normally ascribed to the person. The brain in a vat (i.e. the entire entity of brain + vat), similarly, cannot possibly possess these faculties, since for it to be able to do so would insinuate that the brain (as part of the brain + vat entity) is the agent of such faculties, which is dismissed on the grounds of the mereological fallacy. Therefore, overall, the identification of the brain with the person whose brain it is, is curious and muddled.

5. *The notion that memory is of the past, that it is a collection of stored entities or traces left behind in the physiology of the brain.*
It is interesting to note that Wittgenstein waded against this position in the early twentieth century. He noted, in an unusually vague manner:

> Indeed, I confess, nothing seems more possible to me than that people some day will come to the definite opinion that there is no copy in either the physiological or the nervous systems which corresponds to a *particular* thought, or a *particular* idea, *or memory* [this author's italics]. (*LW I*, §504)

On the basis of this confession, it seems clear that Wittgenstein was at least wary of the prevailing trend of the time which was beginning to suggest (thanks to the historical work of Plato, Locke and Aristotle, which was then developed further by James and Köhler) that memory is neurophysiologically traceable and that it is a collection of stored objects which reside inside the person, mind or brain.

Whereas Plato discussed the idea that memory was to be considered like a piece of wax, Locke considered it more like a storage unit of objects, which amounted to previous experiences and our stored recollections of these experiences. Aristotle first conceived of memory as traceable (i.e. that memory was a trace left behind after an experience took place), and this idea was developed in more detail by James in his work on the principles of psychology, and by Köhler.

Three problems are evident with this view of human memory. First, if the memory is traceable and if memory 'items' are mental objects, then it must have a *physical* location. Secondly, if memories can be recalled at will, then there must be some manner in which they can be interpreted and recognized as past events. Finally, if memory is considered to be a trace or an impression (in the mind or brain) which is left behind after an experience, then there must be some way of accessing it, the only candidate for which must be some form of 'looking inward', such as some form of introspection or 'seeing into the past' (*RPP II*, §592).

However, as Hacker (2014: 5–6) outlines, 'memories need not be of the past – it has to be *acquired* in the past', and further, 'memories [can be stored] neither in the brain nor in the mind. To remember is to *retain* knowledge, not to *store* it.' The difference here is clearly that to retain something, one need not give it a location, whereas to store it, one needs to store it somewhere. Indeed, it is intelligible to suggest, on seeing an old friend, that she has 'retained' her good looks, but it would be a nonsense to say that she had 'stored' them! So too with memories; they are retained, but not stored.

Moreover, memories cannot be isomorphic with traces left inside the physiology after an event has taken place. As Wittgenstein remarks:

> An event leaves a trace in the memory: one sometimes imagines this as if it consisted in the event's having left a trace, an impression, a consequence, in the nervous system. As if one could say: even the nerves have a memory. But then when someone remembered an event, he would have to *infer* it from this impression, this trace. Whatever the event does leave behind in the organism, *it* isn't the memory. (*RPP I*, §220)

In other words, a trace, whatever it is, cannot permit the person who accesses the trace to identify the 'pastness' of the event to which the trace is supposedly isomorphic. Thus, even if the trace captures some event, it remains impossible for the viewer to recognize that this trace is of a past event. The concept of 'pastness', therefore, cannot be inscribed into the physiology. This is why Bax (2012: 36) concludes:

> The role of memories ... cannot be accounted for by stating that they constitute old or retrieved images. ... As Wittgenstein points out, even if images would appear before one's inner eye ... it remains to be explained how one is able to recognize them as representations of something past. Such recognition, namely, requires information that is not contained within the images themselves. ... An image by itself, whether mental or not, does not reveal the date of its production, and talk of retrieved images thus fails to make sense of the very essence of remembering: its allowing us to retrieve or relive things past.

## 8.3 Further Wittgensteinian arguments

There are three other major arguments attributable to Wittgenstein which are directly relevant to this discussion, and are at least worthy of note. The arguments will be summarized, then related to this discussion, afterwards:

6. *The rule-following paradox*
The rule-following argument (*PI*, §§134–242) has been developed many times over,[4] pioneered by Wittgenstein in *The Philosophical Investigations*, and developed further in a raft of secondary literature (Stern 2004: 139–70; Kripke 1982: 7–15; Wright 2001; Malcolm 1986: 154–82; Glock 1996: 323–9; McGinn 1997: 73–112; Kitchen 2014: 101–4) which clearly demonstrates its profound importance to twentieth-century philosophy. Although the argument is deliberated upon throughout Wittgenstein's work, one proposition seems to capture the argument most concisely: 'This was our paradox: no course of action could be determined by a rule, because every course of action can be brought into accord with the rule' (*PI*, §201).

This argument is central to the understanding of rules as insufficient *by themselves* to guide towards their appropriate application. What this suggests, therefore, is that rule-following and rule-governed behaviour are communal in nature, since the correct application of a rule is not captured within the rule itself. As Wittgenstein also remarks in *Remarks on the Foundations of Mathematics*,

---

4     See *Philosophical Investigations*, §§134–242.

'The employment of the word "rule" is interwoven with the employment of the word "same". (*RFM*, MVII, §59). In other words, rule-following behaviour is identified by like-minded individuals doing things *in the same way*.

### 7. The 'beetle in a box' argument

Found in (*PI*, §293), this argument is central to Wittgenstein's rejection of private definitions of concepts like pain, for example. Essentially, Wittgenstein develops this ingenious thought experiment, which demonstrates that if concepts like pain were considered to be private, it would follow that nothing meaningful could ever be defined. In such a private world, the 'beetle' inside the box (akin to the 'thought' inside the brain) which Wittgenstein invokes would be semantically irrelevant, since everyone's beetle could be different, could be the same or there may be nothing to which the beetle refers.

### 8. The private language argument

The argument is generally accepted to be developed over a major section of Wittgenstein's most celebrated piece of work, *Philosophical Investigations*, in particular over §§243–315. There are many core texts in the secondary literature on the topic which offer detailed discussions of this argument (Stern 2004: 171–85; Kripke 1982; Glock 1996: 309–19; McGinn 1997: 113–42), and which this book will not re-examine in any significant detail.

The overarching point Wittgenstein makes in his private language argument, primarily in (*PI*, §243), is to highlight the absurdity of viewing linguistic exchange as private in the sense that it is privately owned and only accessible to its owner. In other words, there can be no possibility of a language which is accessible only to one person, and in this sense is (logically) private.

Indeed, as Wittgenstein rightly remarks, how might such a language even come to be? If, as Wittgenstein argues, the meaning of a language is in its use, then how might a private language attain its meaning? Well, only by its private use. But here we are back to the beetle in the box, where the words which are privately defined become semantically irrelevant, and refer to nothing meaningful. Such a language is useless, even if it is conceivable, which remains highly questionable.

Moreover, since human beings are communal language users, it is beyond the realms of possibility for such a private language to be developed. Indeed, if John, as a communal language user, makes claims of possessing a private language which only he can interpret and use, then it follows that John could translate such a language into his communal language, therefore violating its private nature. It follows, therefore, that any meaningful language cannot be private.

### 8.3.1 Relating these arguments to education

To connect these arguments to education, the author invokes the work of two other major philosophers: Michael Polanyi and Michael Oakeshott, together with work taken from Kitchen (2014: 77–154) where more detailed arguments can be found.

To begin with, consider Polanyi's claim that 'We know more than we can tell' (Polanyi 1983: 4). The articulable aspect of knowledge – that which *can* be said – is what Polanyi would have called 'explicit' knowledge (1983: 4), whereas the part of knowledge which is shown – that which is *more* than we can tell – is akin to what Polanyi labels 'tacit' knowledge.[5]

What seems to be clear, however, is that there are elements of all knowledge which are beyond articulation. That is, these aspects are shown, not told. As learners, when such knowledge is expressed during teaching and experienced during learning, both explicit knowledge (that which can be articulated) and tacit knowledge (that which is developed through close assimilation of a master expert and his *mannerisms and utterances*) are developed.

In such a view of the learning process there is a requirement for a practice and a tradition, which is where meanings are established. On a one-to-one level, between teacher and pupil, this yields Polanyi's explicit–tacit distinctions. Explicit knowledge is imparted from teacher to pupil by didactic-style teaching. However, such explicit knowledge is meaningless – and, in particular, is beyond use – in the absence of the tacit knowledge which is *shown* in the teaching process to the pupil. An ability to use the explicit knowledge is demonstrated by the teacher – the master expert – by developing the tacit knowledge which is beyond articulation. The explicit knowledge is useless without the tacit knowledge, just like the rules of a game are useless to someone who is never shown *how* to play the game.

Here, one is reminded of one of the fundamental undertones of *PI*, captured most succinctly in §43, where Wittgenstein makes the compelling case that *meaning is use*. As Wittgenstein opines, 'For a *large* number of cases … the meaning of a word is its use in the language.' That is to say that syntactic content (words and symbols) are given their meaning by *how* they are used in a language, or more precisely, how these words are used by a community of language users. Without this sociological and anthropological aspect of language (or any *language-game* for that matter), the symbols of the language are bereft of their meaning. Without *use* (shared uses and judgements between like-minded

---

[5]   For a point of clarity, this author parts company with Polanyi in his attempts to establish a 'theory of knowledge' which creates a dichotomy of knowledge 'kinds' to be built inside one another.

individuals) *meaning* drops out of the picture. Inner theories of education (e.g. neuroeducation) cannot account for the element of human participation, and as such can never offer a full description of how learning takes place.

These realizations lead Polanyi to develop a view of education and educational processes which conceive of a form of knowledge development housed inside what he calls a 'fiduciary framework'. At the centre of this fiduciary framework – an educational framework constructed in a baseline trust of one's intellectual superiors – Polanyi emphasizes the importance of traditions, communities and practices which are central to the dissemination of knowledge, both explicit and tacit. The teacher, then, is the representative of these traditions, communities and practices. As Polanyi remarks:

> We must now recognize belief once more as the source of all knowledge. Tacit assent and intellectual passions, the sharing of an idiom and or a cultural heritage, affiliation to a like-minded community: such are the impulses which shape our vision of the nature of things on which we rely for our mastery of things. No intelligence, however critical or original, can operate outside such a fiduciary framework. (1958: 266)

Moreover, As Polanyi (1958: 50) claims, 'Rules of art can be useful, but they do not determine the practice of an art; they are maxims, which can serve as a guide to an art only if they can be integrated into the practical knowledge of the art. They cannot replace the knowledge.'

Mitchell (2006: 63) develops this claim about the progression of practical knowledge:

> Practical knowledge precedes the knowledge of rules … one must possess a degree of practical knowledge in order to properly apply rules. … One acquires practical knowledge through doing … how can one practice an art if one does not yet know how to do so? The answer lies in submission to an authority in the manner of an apprentice. We learn by example.

Indeed as Polanyi (1958: 53) also argues:

> To learn by example is to submit to authority. You follow your master because you trust his manner of doing things even when you cannot analyse and account in detail for its effectiveness. By watching the master and emulating his efforts in the presence of his example, the apprentice unconsciously picks up the rules of the art, including those which are not explicitly known to the master himself. These hidden rules can be assimilated only by a person who surrenders himself to that extent uncritically to the imitation of another.

One can see here the connection between rules and their application. Knowing a rule (syntactic, explicit) is insufficient for the 'art' of knowing. Knowing its application, however, equips one to go beyond the information given. This is why Polanyi places so much emphasis on his fiduciary framework, because inside such a framework the learner can assimilate his master and develop an ability to proceed beyond the information (explicit) given to him in his learning. This process, however, depends on the pupil fostering the ideas of his teacher – in some cases unconsciously or without question – and in this case the teacher is representing the community, tradition and practice from which he comes.

In a similar manner, and indeed more overtly applicable to the educational aspect of this discussion, Oakeshott talks of the distinction between information and judgement. According to Oakeshott, information 'consists of facts, specific intellectual artifacts (often arranged in sets or bunches). It is impersonal (not a matter of opinion). Most of it is accepted on authority, and it is to be found in directories, manuals, textbooks and encyclopedias' (1989: 45).

Moreover, Oakeshott (1989: 46) suggests that 'the importance of information lies in its provision of rules or rulelike propositions relating to abilities'. In essence, therefore, information is the component of knowledge that provides the 'knower' with the rules which govern the ability to act. Indeed, as Oakeshott (1989: 46) opines, 'Every ability has its rules, and they are contained in that component of knowledge we call information. This is clearly the case with mathematical or chemical formulae.'

This definition of information clearly relates to the syntactic content of knowledge; that is, the rules and rule-like propositions which are static and inert. Information is insufficient on its own to construct the knowledge edifice, though any attempts at trying to develop knowledge which does not encompass information and facts is, according to Oakeshott's definition, meaningless. However, 'before any concrete skill or ability can appear, information must be partnered by "judgement", "knowing *how*" must be added to the "knowing *what*" of information' (Oakeshott 1989: 49). This is what is required to give the information 'part' of knowledge its meaning: judgement. Oakeshott (1989: 49) defines the concept of 'judgement' as follows:

> The tacit or implicit component of knowledge, the ingredient which is not merely unspecified in propositions, but is unspecifiable in propositions. It is the component of knowledge which does not appear in the form of rules and which, therefore, cannot be resolved into information or itemized in the manner characteristic of information.

Judgement, therefore, is the semantic content of knowledge. It resides outside of the confines of syntactic content, which is explicit (to use Polanyi's language) and articulable. The rules and their meaning are not of the same form; a rule does not guide to its appropriate use, and an understanding of a rule is not contained within it. An ability to act in agreement[6] with the rule requires judgement, and this judgement resides in the outer, public world, not in the inner, subjective world. That is, judgement is established within a community of people who have had the same training, and whose familiarity with the rules of the practice is a shared common core of ideals and beliefs. Since 'agreement' is not an individual term, the ability to interpret and use a rule clearly depends on a social aspect of rule-following which can never be captured by an examination of inner facts alone. Indeed, as Malcolm (1986: 156, original emphasis) observes, 'If there was no *we* – if there was no agreement among those who have had the same training, as to what are the correct steps in particular cases when following a rule – then there would be no *wrong steps*, or indeed any *right steps*.'

Moreover, as Wittgenstein questions on the nature of rule-following:

> How can I follow a rule, when after all whatever I do can be interpreted as following it? (*RFM*, VI, §38)

Wittgenstein answers thus:

> It is true that *anything* can be somehow justified. But the phenomenon of language is based on regularity, on agreement in action.
>
> Here it is of the greatest importance that all or the enormous majority of us agree in certain things. I can, e.g., be quite sure that the colour of this object will be called 'green' by far the most of the human beings who see it.
>
> …We say that, in order to communicate, people must agree with one another about the meanings of words. But the criterion for this agreement is not just agreement with reference to definitions, e.g., ostensive definitions – but *also* an

---

[6]  It is important to distinguish between acting in accord with a rule and acting in agreement with a rule. 'Agreement' is a communal/comparative term, which stipulates that there is an anthropological and sociological aspect to following a rule *in agreement*. In other words, 'agreement' is a closely related concept to the concepts 'community' and 'correctness'. As such, acting in agreement with a rule requires that one is able to distinguish between the correct and the incorrect application of the rule. Acting in accord with a rule, however, need not adhere to any such restrictions. It is entirely possible any course of action to be brought into *accord* with a rule (*PI*, §201), since *under some interpretation* of the rule, every course of action accords with the rule. It is only when one introduces the concept of 'correctness' into the picture that one simultaneously introduces the ability to act in agreement with the rule. Indeed, as Malcolm (1986: 156) observes: 'If there was no *we* – if there was no agreement among those who have had the same training, as to what are the correct steps in particular cases when following a rule – then there would be no *wrong steps*, or indeed any *right steps*.'

agreement in judgements. It is essential for communication that we agree in a large number of judgements. (*RFM*, VI, §39, original emphasis)

Oakeshott (1989: 50) summarizes:

What is required in addition to information is knowledge which enables us to interpret it, to decide upon its relevance, to recognize what rule to apply and to discover what action permitted by the rule should, in the circumstances, be performed; knowledge (in short) capable of carrying us across those wide open spaces, to be found in every ability, where no rule runs.

In other words, the answer to Wittgenstein's curious question on the nature of rule-following in *RFM*, VI, §38 is found in the realization that the ability to follow a rule in agreement – which this author contends is a primitive necessary condition for the definition of learning – is captured in an agreement of *judgements*, which is the facet of knowledge which equips the learner to *do*; that is, to *act* in such a manner which is seen to be *correct*. Such correctness is not an internal facet of the human being, rather a collective feature of the human race which no internal educational theory can ever capture.

Part Four

# Intrinsic and Relational Models of Education: Unifying the Philosophy of Mind and the Philosophy of Quantum Physics

# Intrinsic and Relational Models of Education

## 9.1  Introduction to Part 4

The prevalent theme of this and indeed the subsequent chapters will be the following: the *within* and the *without* give way for the *between*. By unifying Wittgenstein's philosophical ideas together with Bohr's philosophical views on quantum physics – an unlikely combination, granted – this final part of the book aims to replace the Cartesian–Newtonian paradigm of psychology and education with a Wittgensteinian–Bohrian model, more suited to modern-day developments within philosophy and science.

## 9.2  Philosophical interlude

In his PhD thesis, *Tractatus Logico-Philosophicus*, Wittgenstein outlined his belief, 'The limits of my language mean the limits of my world' (§5.6). Furthermore, in his author's preface of the publication version of the same text, Wittgenstein outlines that the aim of the *Tractatus* 'is to draw a limit to thought, or rather – not to thought, but to the expression of thoughts' (*TLP*, 68).

What this offers is one of the first philosophical arguments that Wittgenstein puts forward within the body of his work to outline the importance of communication; and, indeed, the significance of the *relational,* communitarian nature of such communication. So much was this significance, that Wittgenstein restricts his entire world within the limits of his language. Meaningful linguistic exchange requires more than the language; it also requires language-games, which themselves are formed inside a communal, sociological and anthropological framework; language is 'part of an activity, or a form-of-life' (*PI*, §23).

Despite the fact that Wittgenstein moved away from much of his work in the *Tractatus* in writing the *Investigations,* his point here is a simple one: that which one cannot express in an *accepted* form of language is considered to be beyond the limits of one's language, and thus, by extension, is beyond the limits of worldly concepts. Language – rather, communication – therefore lies at the heart of everything meaningful. That is not to say, however, that that which cannot be said is non-existent: 'There are, indeed, things that cannot be put into words. They make themselves manifest. They are what is mystical' (*TLP*, §6.522). However, Wittgenstein suggests that these unarticulated 'things' are taken as a baseline; that they are mystical and that they are simply found in action, not in articulated linguistic exchange.

The core ideas from Part 3 of this book can therefore be related now to one prevailing idea: namely, that meaningful action, communication and behaviour are founded in the communal, anthropological and sociological aspect of so-called language-games, which are embedded into a form of life and a world-picture which is restricted by the limits of this language.

This realization is one of the important cornerstones of the connection between Wittgenstein's philosophy of mind and Niels Bohr's philosophy of physics. For example, what Wittgenstein regarded as *language-games*, Bohr would have described as *unambiguous communication*. Moreover, taking Wittgenstein's philosophy as an argument in support of what might be called *anthropological holism*, that is, the development of shared meanings over time, there are clear and obvious connections to what Bohr often called 'inter-subjectivity' or 'weak objectivity'.

One pre-emptive comment before this remit is fulfilled: it might often be argued that education does not need philosophy, and it might also be argued that the connections between the respective philosophies of physics and of education are more than a little curious. It is this author's view, however, that the lynchpin between these respective philosophies is found in an examination of the philosophy of psychology. Indeed, it will become clearer later in this chapter that there is credence in examining the philosophies of physics and psychology with respect to one another; this has, in fact, been one of the core discussions from within the quantum physics movement, with contributions from Einstein, Bohr, Heisenberg, Oppenheimer and Stapp – among many others – to the debate.

It is interesting to note what Oppenheimer warned in 1955, during the height of the quantum revolution within physics that psychology can no longer depend on a philosophical model of Newtonian determinism, when physics had begun to move in another direction: 'It seems to me that the worst of all possible misunderstandings would be that psychology be influenced to model itself after a physics which is not there anymore, which is quite outdated.' Similarly,

Stapp outlined that 'while psychology has been moving towards the mechanical concepts of nineteenth-century physics, physics itself has moved in just the opposite direction' (1993: 192).

Furthermore, quantum pioneer Niels Bohr outlines his support for the link between physics and psychology – and by this author's extension, for education also – claiming that there exists 'an epistemological argument common to both fields', going on to stress the importance 'to see how far the solution of relatively simple physical problems may be helpful in clarifying the more intricate psychological questions with which human life confronts us, and which anthropologists and ethnologists so often meet' (Bohr 2010: 27). This author contends that educationalists should be added to Bohr's original list. Wick (1995: 185) also notes of Bohr's work in quantum physics that 'There seems to be a reference to psychology in every exposition Bohr wrote on complementarity.' Indeed, as Bohr himself argues, there is a 'close analogy between the situation as regards the analysis of atomic phenomena … and the characteristic features of the problem of observation in human psychology' (2010: 27).

It is also interesting to note that Bohr was aware of the continued questionable commitment of the psychological sciences to the classical/Newtonian paradigm of physics even so far back as 1938, made clear in his keynote lecture to the International Congress of Anthropological and Ethnological Sciences in Copenhagen, entitled 'Natural Philosophy and Human Cultures' (2010: 27).

It is these connections and concerns of which Bohr speaks that this author will examine over the final part of this book. Furthermore, this preamble serves as sufficient evidence to proceed on the basis that the psychological sciences, quantum physics, and indeed education have a shared philosophical underpinning which requires further examination. Moreover, it seems that these connections are found in the conceptual and methodological paradigms for measurement in physics, psychology and education alike.

The idea is that the model of measurement which is used in quantum physics – relational measurement – when extended to psychological phenomena in general and educational attributes and predicates in particular (e.g. learn, think, understand, intelligence, reading ability and mathematical ability) gives rise to a new, non-Newtonian, non-classical model of psychology and education which fits neatly inside the Wittgensteinian (*ipso facto* non-Cartesian) philosophy which this author has outlined in Part 3 of this book. That is, it will be shown that this relational model of education accounts for an infused inner/outer picture, with the practices, customs and traditions playing an indispensable role in the description and ascription of all educational phenomena.

## 9.3  Relational and intrinsic attributes and abilities

Much of Part 3 of this book could be taken as a preamble to the furtherance of the so-called *relational model* of education, which this author will subsequently focus on developing. Before this can be done, however, there is a requirement to provide a basic philosophical definition of relational attributes, as well as a definition of the counterpart to relational attributes, so-called *intrinsic attributes.*

### 9.3.1  Definitions

At this juncture of the discussion it is only necessary to give fairly primitive definitions of these terms, with the view to seeing, in a general sense, what the restrictions of intrinsic attributes are and what the advantages of relational attributes are in this author's descriptive model for educational attributes in particular. The notion of relational attributes will be discussed further in the book, with particular reference to the quantum physics ideas surrounding relationalism.

An *intrinsic attribute* is an attribute which is a feature of an object or entity, existing independently of the act of measurement. Consequently, intrinsic attributes can be measured using a process of 'checking up'. For example, a person's height is an intrinsic attribute of the person. When the person's height is not being measured, it still *makes sense* to ascribe height to him. In this sense, height is intrinsic because the concept of 'height' can still be ascribed to a person even in the absence of measurement. To say, for example, that 'John only has height when he is measured' would be absurd! Such claims are not in keeping with our understanding of the world. Moreover, the definition of the attribute 'height' is independent of a measuring instrument. To be clear, in order to specify *a particular* height a measuring instrument is required, for example, a tape measure, a ruler or a clinometer. However, in the absence of the measuring instrument, the attribute 'height' can still be meaningfully said to exist. The *removal of the measuring instrument* in no way affects the definition nor the ascription of the attribute 'height'.

The role of the measuring instrument, therefore, in measuring intrinsic attributes is one for the reduction of the uncertainty surrounding the *precision* of the attribute in question. The attribute itself, however, does not *depend* on the measuring instrument for its definition or ascription.

A *relational attribute* in contrast is an attribute which is *dependent* on the measuring process for its meaningful ascription. That is, in the absence of the measuring process one cannot meaningfully speak of the existence of the attribute in question. It is not, as a matter of fact, that the attribute *does not exist* when measurement is postponed, as opposed to the attribute having no meaning when separated from the process used to measure it. When measured, relational attributes take on definite properties; when unmeasured, the attributes in question are suspended in a state of indeterminism, taking no particular form. These attributes are thus called 'relational' because their definition and ascription are a *joint property* of the object being measured and the measuring device and process being used. If one removes the measuring instrument from the definition thus ceasing the measuring process, one loses the attribute also, insomuch as the attribute takes on no definite properties – and, by consequence, takes on *all possible* properties – whenever it is not being measured. Only when it is measured can it be meaningfully ascribed with definite properties and qualities.

Take, for example, the attribute 'reading ability'. One comes to appreciate the profoundness of this definition of relational attributes when one ponders: 'What might it mean for a person to possess an ability to read, and more importantly, what would it mean to *ascribe* reading ability to a person, in the absence of *a reading test* (which in this case serves as the measuring instrument) or indeed in the absence of *the practice* of reading (which prevails as the benchmark for any measuring process which seeks to "capture" reading ability as an attribute)?'

Alternatively, one might ask: 'How is the attribute "reading ability" ascribed to a person, and in what way does this make the "ability to read" a relational attribute?'

It seems intuitive to assume that an ability to read is a feature of the person in question. That is, it seems perfectly reasonable to hold that a person's ability to read remains 'part' of him, even when he is not reading. Therefore, so it would seem, when he reads, he is simply demonstrating what is a pre-existing *feature of him*. But the picture here is a little different. Wittgenstein notes that these concepts are not so clearly defined:

> We also say, 'Since yesterday I have understood this word'. 'Uninterruptedly', though? – To be sure, one can speak of an interruption of understanding.
>
>   ...
>
>   What if one asked: When *can* you play chess? All the time? Or just while you are making a move? And the whole of chess during each move? (*PI*, §149 (a) and (b))

This is the essence of a relational attribute, and Wittgenstein is using several analogies here to show precisely what a relational attribute is. Indeed, an ability to play chess, for example, is shown here to be measured in relation to the prevailing practice of 'playing chess'. Wittgenstein is questioning here whether one *can* meaningfully ascribe an ability to 'play chess' when one is not engaged in the act of actually playing it. Similarly, when one comes to 'understand' a word, can one be said to understand the word even in the cases when one is not demonstrating one's understanding? It seems that, in these cases, the attributes in question are meaningfully ascribable only in the instances when they are measured.

The defining difference here is the need for some sort of comparison to and engagement with a prevailing practice in order to *make manifest* the attribute in question. Therefore, the attribute – relational in nature – is a joint property of the person who makes it manifest and the pre-existing practice which is engaged with and compared to in order to make the measurement. If the practice – the measuring instrument – is removed, then the manifestation is no longer possible, and the attribute becomes undefinable and is beyond manifestation. Furthermore, if the measuring instrument is changed then the interaction between the measured object and the measuring instrument also changes, and, as such the definite properties of what is being measured also change.

If one considers again the notion of intrinsic attributes, and the attribute 'reading ability' it becomes clear why reading ability cannot be regarded as an intrinsic attribute. Consider how the intrinsic attribute theorist might try to measure – or 'check up' on – the reading ability of a person. The observer would equip the person with an item of reading material and give the instruction 'read', upon which the observed person would read the item of prose. Depending on the difficulty of the item of prose, and a host of other observational factors – such as reading intensity, pace and ability to recognize context – the observer would then ascribe reading ability to the observed person. There seems, on first viewing, to be no apparent difficulty with the assumption that the observed person had a pre-existing ability to read, and the item of prose was simply used to examine that ability.

The explanation might go a little further – in keeping with the neuroscience model, for example – to suggest that the various parts of the frontal and temporal lobes are where the reading ability resides. Neuroscientific studies *seem* to show this. The experiment therefore, might entail an examination of the activity which takes place in these areas of the brain while the observed person reads the prose.

This entire experiment assumes that reading ability is a feature of the person, it is embedded in the person, and when it is measured the measurement simply captures an attribute which already existed and is now shown in the observation on the neural scan. However, there is one glaring problem: the nature of the attribute which is being measured – reading ability – clearly depends on the measuring instrument used to measure it for its meaningful ascription. That is, any experiment or observation which attempts to 'capture' someone's ability to read will always depend on the measuring instrument used to measure the attribute. The measurement is a joint property of the observed entity and the instrument used to measure it. If we ask Sarah to read *Green Eggs and Ham*, we might find that she is a fluent reader, but if we offer her a piece of prose from Dostoyevsky's *Crime and Punishment* we might find that she cannot come to terms with the more complex language. If we change the instrument of measurement, we change the nature of what is being measured. Moreover, the entire ascription of reading ability depends on a pre-existing *practice* of reading. *It is not the individual attribute 'reading ability' which pre-exists, as opposed to the practice of reading,* which is a public possession, not a private one. In the absence of the practice of reading – the measuring instrument – the measurement of the ability is no longer possible. Therefore, the ability to read cannot be meaningfully said to pre-exist the act of measurement; that is, it cannot be regarded as *intrinsic* to the observed person.

The role of measurement seems to be the defining difference between intrinsic and relational attributes. For an intrinsic attribute, measurement of the attribute in question is used to capture, more or less precisely, some pre-existing state of the object under observation. In contrast, for a relational attribute, the measurement cannot be meaningfully said to capture some pre-existing state; rather the attribute which is captured can only be meaningfully said to exist when it interacts with the device or instrument used to measure it. If one removes the measuring device, one loses the attribute. If one changes the measuring device, one changes the measurement, and thus also changes the interaction between the observed and the observer, as well as the measured attribute and the measuring device.

## 9.3.2 Educational attributes: Relational or intrinsic?

On a more general level, the appeal of a relational attributes paradigm for educational attributes, such as learning, thinking, understanding, mathematical ability and intelligence, over an intrinsic attributes paradigm becomes clear in the context of these definitions. It will henceforth be the remit of this author in this part of the book to set down a relational attributes model of description and ascription for educational attributes. The foundation for the appeal to a relational attributes model becomes even clearer when set against the backdrop of quantum physics, which will also be discussed in subsequent chapters.

These primitive definitions of intrinsic and relational attributes connect also to philosophical ideas outlined earlier in this book, further making clear the appeal of a relational attributes model. To begin with, relational attributes are governed by first-person/ third-person asymmetry and they adhere to all the intricacies of asymmetrical ascriptions. Intrinsic attributes, however, are symmetrical in their ascription. As such, there is a basis in the most primitive of forms to appeal to relationalism in place of intrinsic ascriptions when dealing with psychological attributes, which this author has shown to be governed by asymmetry.

Secondly, the constitutive uncertainty of psychological and educational attributes, which by definition is irreducible, is captured by relational attributes, but not by intrinsic attributes. Indeed, with respect to educational attributes for example, within the relational paradigm, the uncertainty of third-person ascriptions – when such uncertainty exists – is not so much a deficiency of the measuring instrument, as it is a part of the relational nature of the attribute in question. That is, if one regards educational attributes as relational in nature, then one ceases to be concerned by the picture of the uncertainty of the attributes when they seem to be 'hidden from view' since in the relational model, the concept of a 'hidden' attribute, in the absence of measurement, is unintelligible. The attribute is either made manifest by the measuring instrument – in which case it takes on a particular form – or it is not – in which case it takes on no (and, by extension, all possible) meaningful attributes. Nothing is hidden in relational models; attributes are either manifest or non-manifest depending on the use or non-use of a measuring instrument. It is only when one attempts to reconcile educational attributes with an intrinsic attribute model that one faces the question of 'hidden' or not. Thus, a commitment to searching for the hidden (intrinsic) attribute – however sophisticated the techniques used – leads to attempts to resolve the irreducible uncertainty of the attributes in question, which will inevitably lead only to failure.

Furthermore, the relational attributes paradigm is more suited to account for what this author has labelled 'the infused inner/outer relation', in contrast to the intrinsic model which attempts, by definition, to detach the inner from the outer, with greater emphasis naturally placed on the checking-up of the inner only. Built into the relational model is the idea that the object under observation and the observing device are inseparable in the process of meaningfully ascribing attributes to the observed object. The attribute is a feature of both the object – the inner – and the measuring device – the outer. In an intrinsic model, the observed object is separable from the measuring instrument, and the attributes of the object can be meaningfully ascribed independently of the measuring instrument. The intrinsic model therefore cannot capture the infused inner/outer picture of the ascription of psychological and educational attributes since it treats the observer and the observed as independent, not infused. The separation of the inner and the outer upon which the intrinsic model is based thus makes the model unsuitable for the observation and description of educational attributes.

The attraction, therefore, of the relational model for educational attributes and phenomena is captured in what this author has claimed previously: the *within* and the *without* give way for the *between*. When observing and describing educational attributes, it is not the inner nor the outer which is of importance, as opposed to being the *relation* between the inner and the outer. It is the *interaction* between the observed and the observer which leads to a suitable model of ascription and description of the phenomena in question, and the only model which accounts for this *holism* and *inseparability* is the relational attributes model. The educational paradigm, therefore, must be rooted in relationalism.

### 9.3.3 Two alternative worlds: The importance of communication – the 'between' of relationalism

The eminent philosopher, Hilary Putnam once famously remarked that 'meanings ain't in the head'. He developed a clever thought experiment to make a compelling (yet perhaps flawed) argument that semantic internalism was a flawed doctrine, which ought to give way to definitions and meanings being externally fixed.

Consider a revamped version of this argument, which we might call 'The Two Alternative Worlds' thought experiment. The point of this experiment is to demonstrate the absurdity of both the 'within' and the 'without', which this author contests should both give way for the 'between' of relationalism.

1. *Alternative world 1: Introvert*
Consider first alternative world '1' which, for the sake of the argument will be called 'Introvert'. In this world, everyone is born in such a way that no form

of communication between people is possible. No language is developed, no utterances are possible and there is no communally accepted form of interacting with one another. There are no forms of articulation, it is not possible to express pain, hunger, tiredness, joy, intellect, thinking, ruminating or learning in any manner which is akin to these concepts on earth.

On Introvert, what does it mean to be in pain as we would know it on earth? Is it an inner process which is privately owned by he who experiences it? Is John's experience the same as Sarah's? Are these questions even intelligible, especially given that they can never be tested? Moreover, isn't it conceivable on Introvert that there are *no* experiences of any kind – even 'inner' experiences? Perhaps nothing at all goes on behind the eyes.

The point here is a simple one: in the absence of communal forms of communication and linguistic exchange, do not only so-called 'outer' experiences drop out of the picture, but the intelligibility of 'inner' experiences disappears as well.

This is the first aspect of why relationalism is required, and the importance of the community and the practice respectively in what might be called 'educational relationalism' is highlighted. Indeed, in the absence of agreed forms of communication, and in the absence of a practice which the community cultivate and work within, the semantic content disappears from our concepts and experiences. On Introvert, communication is non-existent, and practices are not formed. As a consequence, it becomes impossible, and indeed unintelligible for the inhabitants of Introvert to engage in any form of meaningful activity, either with others, or, in fact, in a 'private' moment. On Introvert, the only activities which are permitted are private ones; but, paradoxically, the prohibition of public experience and communal activity and engagement means that even the so-called private experiences are meaningless. They are in fact, like Wittgenstein's beetle: they become irrelevant to a description of the world-picture on Introvert.

In a world where psychological and educational attributes are governed by relationalism, however, these problems are avoided. Abilities are measured against the prevailing practices and traditions, and their definitions are founded on the relation between the practice and the interaction with the practice by those who engage with it.

Returning from the thought experiment for a moment, it is clear why relationalism captures a necessary component of the fullest description of educational attributes, which is absent in the Introvert example. That is, the interaction between the person and the practice – the relation – is what offers a full picture of what it means to have reading ability, mathematical ability and so on. This dispenses with the importance of the 'within' *only*, and shows why the 'between' is the focus of the description of nature.

## 2. *Alternative world 2: Extrovert*

Similarly, consider alternative world '2' which will be called 'Extrovert'. On Extrovert, there is no concept of reflection or thought, only actions, reactions and outward behaviour. Nothing motivates behaviour, and there is no concept of intention. *All* actions are capricious, born out of instinct. As such if Jim and Pauline act in a similar manner on Extrovert, they cannot be said to act for any reason in particular, and so their actions are taken at face value only. If Jim and Pauline both withdraw £1000 from the bank, they both do so in caprice, and nothing can be said of their respective intentions for the money, until action is taken, after which we can observe that Jim uses his £1000 to pay off his gambling debts, whereas Pauline takes her mother on a long overdue holiday.

How absurd is a world in which intention drops out from the picture? How unintelligible is it that action – that is, all action – is born out of caprice? Intention – that is, the act of intending to act in such-and-such a way – is an intricate concept. To be clear, there are occasions when our intention to act is seemingly entirely capricious in nature. There are indeed instances when no discernible source can be traced for one acting in such-and-such a manner. Nonetheless, there are a great many cases when action is intended only after thought, reflection on one's thoughts and considerations of one's potential actions. All of these activities depend on one grasping the potential benefits and pitfalls of a particular course of action, and thus setting about to extinguish and avoid the pitfalls, and pursue the benefits.[1] One moment's pause for reflection will highlight that what binds this picture together is the concept of *understanding*, which itself is predicated on *meaningful action*.

On Extrovert, behaviours are not rule governed and they are born entirely outside of the bounds of any kind of order. Any two people might be found to carry out the same actions, and even the omniscient being looking for their respective motives would find no reason to delineate between their actions, or the reasons for their actions. Their intentions are absent.

So too is their understanding of their actions. On Extrovert, there is *only a practice*, but there is no meaningful interaction with it. The child who reads *Green Eggs and Ham* and *Crime and Punishment* does so bereft of motivation,

---

[1] This, however, should not be misunderstood as an espousal of a 'causal' model of description.

engagement or understanding. The boy who solves quadratic equations in mathematics does so only insofar as he is shown, and lacks the *judgement* to solve as-yet-unseen problems. These are the consequences of prohibiting him from thinking about what he does. What might be described here as 'social conditioning' – a behaviourist concept – shows only a modicum of detail in a tapestry of what it *means* to know 'what' and to know 'how'. On Extrovert, practices and traditions are fixed, inert and unchangeable. Nothing can be contributed to the prevailing communal activities, and no new concepts can be developed.

## 9.3.4 Further contributions from Polanyi, Ryle, Oakeshott and Wittgenstein

Polanyi, Ryle, Oakeshott and Wittgenstein have all highlighted the complications with this view of the world (although Ryle's commitment to a mutated form of psychological/logical behaviourism makes his contestations seem out of place). Some of these ideas were discussed in Chapter 8.

1. *Michael Polanyi*
Polanyi dichotomizes the 'art' of knowing into tacit and explicit knowledge; that which could be articulated, and that which was beyond articulation (Polanyi 1983: 4). In doing so, Polanyi makes it clear that there are aspects of knowing which entail public engagement (explicit knowledge) and private reflection and imitation of one's master and his ways of working (tacit knowledge). The interaction between the individual and the practice is captured in Polanyi's so-called fiduciary framework (Polanyi 2009: 266). In the development of these ideas, Polanyi makes the case that there is a significant role to be played in the nature of knowing by the practices, traditions and communities (Kitchen 2014: 84–90).

This author contends that this view of the art of knowing – and indeed, the development of any intellectual facility, ability or capacity which is of interest to education – is captured in the relational educational paradigm. In the relation between subject and object, as this author has defined earlier, it is the interaction between what Polanyi has categorized as 'tacit' and 'explicit' which relationalism captures. Explicit knowledge, as Polanyi defines it, is found in the relational attributes paradigm in the guise of the prevailing practice. It exists *explicitly*, and it is cultivated and disseminated in an overtly public manner. It is, in fact, a public possession, to be shared by a community of like-minded masters.

Tacit knowledge, however, is found in acts of personal engagement with the practice. It is found in one's imitation of a master who is accustomed in the ways of the community which he represents, and the tradition from which he comes. The relation between the subject and the object, the first- and the third-persons, the measured and the measuring instrument, the person and the practice is synonymous with the interaction between tacit and explicit 'knowing'.

## 2. *Michael Oakeshott*

In a similar manner, as this author has outlined already, Oakeshott categorizes knowledge as an equal partnership between information and judgement (Oakeshott 2001: 45; Kitchen 2014: 124–42; cf. Chapter 8). Individually, information and judgement serve as the 'without' and the 'within' of the interaction between the person and the practice. Thus, in keeping with the relational educational model, and with ideas articulated earlier in this book, one comes to appreciate the notion of subverting the 'within' and the 'without', in place of *the relation between* information and judgement.

According to the definition, relational attributes are measured in such a manner that the attribute becomes a joint property of the measured object and the device used to measure it.

Extending this to Oakeshott's language regarding knowledge, one finds that this 'jointness' is found in the interaction between 'information' and 'judgement'. Information – the mere brute facts and figures of knowledge – is the public possession, residing in a community who adopt them. To interact with information in any meaningful manner, however, requires insight and *judgement*, which is, in some sense, a personal ritual which is developed by closely imitating the tacit workings of one's masters. This relation between information and judgement is captured in the relational educational paradigm, where one's grasp of information is infused with one's ability to judge as to its appropriate, that is, correct use.

## 3. *Gilbert Ryle*

In *The Concept of Mind*, Gilbert Ryle also outlines the distinction between what he calls 'knowing that' and 'knowing how'. In fact, Ryle's discussion of these ideas defeats what is now famously known as the 'intellectualist legend' (Ryle 2009: 16) or 'the ghost in the machine'. Unsurprisingly, this was Ryle's critique of Cartesian dualism, which this author has argued against repeatedly over the course of this book.

Ryle's famous distinction between 'knowing that' and 'knowing how', as well as his subsequent argument that there lacks a causal connection between thoughts, intentions and actions brings him dangerously close to behaviourism:

> When we describe people as exercising qualities of mind, we are not referring to occult episodes of which their overt acts and utterances are effects; we are referring to those overt acts and utterances themselves. (Ryle 2009: 14)

Ryle goes on to outline various important points which are applicable to this discussion, in particular with reference to the unity and infusion which exists between the 'knowing how' and 'knowing that'.

These realizations are important for the relational educational model of description of educational attributes because they further strengthen the case that abilities to 'do' activities, such as reading and mathematics, in a meaningful manner are not a dichotomy of separable 'inner' and 'outer' concepts.

Ryle (2009: 15) attributes this mistaken conception of abilities and intellectual capacities to a confusion between intellectual and mental conduct, claiming that 'both philosophers and laymen tend to treat intellectual operations as the core of mental conduct.' As a consequence, this manifests as a commitment to 'the idea that the capacity to attain knowledge of truths was the defining property of a mind' (Ryle 2009: 15).

This is precisely what Ryle categorizes as the 'intellectualist legend' or 'the dogma of the ghost in the machine' (2009: 16). It is, in fact, a residue of Cartesian dualism, the inherent problems of which this author has already strongly contested. As Ryle outlines:

> People tend to identify their minds with the 'place' where they conduct their secret thoughts. They even come to suppose that there is a special mystery about how we publish our thoughts instead of realising that we employ a special artifice to keep them to ourselves. (2009: 16)

This identification, according to Ryle, stems from the fact that we are drawn towards a picture of things which separates the 'theory' from the 'practice', the knowing from the doing. However, such divorce tends towards this separation of the relata in question; relata which ought not to be split, and which the relational model makes no effort to pull apart.

It is this author's view that relationalism is in fact, on the contrary, committed to a full description of psychological (and educational) phenomena by ensuring the precise opposite of what Cartesianism attempts; namely, to consider the measured and the measuring device as an indivisible whole.

In contrasting how a child reads Dr Seuss in comparison to reading Dostoyevsky, for example, one realizes that in ascribing reading ability, it is not the inner 'within' nor the outer 'without' which gives a description of the relata, as opposed to an *infused, relational description* which takes account of the 'knowing that' and the 'knowing how' of reading. It might be possible for one to acquire or develop a cornucopia of reading 'theory', the facts of grammar and the rules of syntax, but only when one practises the activity of reading, and when one engages with the prevailing practice of reading, can one be said to know *how* to read. Inner examinations cannot show this; relational ascriptions capture it entirely.

The relational model also has the advantage of being able to offer a full description, in a manner which *communicates unambiguously*[2] about psychological attributes, about what Ryle (2009: 17) defines as 'intelligence':

> To be intelligent is not merely to satisfy criteria, but to apply them; to regulate one's actions and not merely to be well regulated. A person's performance is described as careful or skilful, if in his operations he is ready to detect and correct lapses, to repeat and improve upon successes, to profit from examples of others and so forth. He applies criteria performing critically, that is, in trying to get things right.

In other words, 'when we describe a performance as intelligent this does not entail a double operation of considering and executing' (Ryle 2009: 18).

The emphasis on the problems of viewing intelligent acts as a 'double operation' here highlights the inseparability of knowing and doing. Relationalism captures this inseparability precisely. Intrinsic models of description cannot, given their focus on a pre-existing state which is supposedly stored somewhere in the locality of the person. Indeed, in the intrinsic model, where psychological attributes are assumed to be intrinsic to the person being measured, one is automatically committed to an inner–outer causal picture steeped in Cartesian dualism.

The model is predicated, tacitly or otherwise, on the view that the attributes which are measured are stored ('inside'?), and thus actions are only made manifest after some prior process which takes place privately. Ryle is saying here that this process – which intrinsic models must take as a double operation – is, in fact, one single, inseparable action. Ryle's defeat of the intellectualist legend and the dogma of the ghost in the machine, therefore, serves equally well as an

---

[2]   This term – 'unambiguous communication' – will be returned to in Chapter 11.

argument *for* a relational attributes model of description for all psychological (and educational) phenomena.[3]

It is important to note the main argument against this intellectualist legend, which Ryle (2009) puts forward as an infinite regress. The idea is that if all intellectual endeavours are dichotomized into the exercise of some mental faculty which is distinct and separable from the active or behavioural output, then meaningful action becomes embroiled in an infinite regress:

> The crucial objection to the intellectualist legend is this. The consideration of propositions is itself an operation the execution of which can be more or less intelligent, less or more stupid. But if, for any operation to be intelligently executed, a prior theoretical operation had first to be performed and performed intelligently, it would be a logical impossibility for anyone even to break into the circle. (Ryle 2009: 19)

Ryle also contends:

> According to the legend, whenever an agent does anything intelligently, his act is preceded and steered by another internal act of considering a regulative proposition appropriate to his practical problem. But what makes him consider one maxim which is appropriate rather than any of the thousands which are not? (Ryle 2009: 19)

What Ryle is claiming here is that if one maintains a commitment to the view of psychological and intentional (and by extension, educational) phenomena which posits that there is an 'inner-causes-outer' cycle then one reaches a regress.

Moreover, the notion that intelligent (outer) behaviour can only follow from intelligent ruminations, which presumably take place inside the mind, induces a profound problem; namely, how can one choose an appropriate course of action to carry out in an intelligent manner, if all intelligent behaviour is itself predicated on preceding intelligent behaviour? The intellectualist legend serves only to internalize a problem which was unresolvable in the outer.

However, the stark realization here is that this problem only manifests itself when one maintains the commitment to the profound dualism of the intellectualist legend, which seeks to dichotomize the inner and the outer. Relationalism pre-emptively dodges the curve-ball of the intellectualist legend by maintaining a holism – contrasted to a dualism – between the measured

---

[3] Ryle invokes the example of intelligence to highlight his paradigm; but the arguments can be extended analogically, without loss of generality, to all forms of psychological phenomena.

and the measuring device, between the person and their interaction with the prevailing practice, between the subject and the object.

Ryle concludes therefore that it is essential to depart from this profound error which is embedded into the intellectualist legend.

> To put it quite generally, the absurd assumption by the intellectualist legend is this, that a performance of any sort inherits all its title to intelligence from some anterior internal operation of planning what to do.
>
> ...
>
> [However] When I do something intelligently, i.e. thinking what I am doing, I am doing one thing and not two. (Ryle 2009: 20)

This, indeed, is the basis for the adoption of the relational model for the unambiguous description and ascription of educational attributes. Despite the fact that over the course of his arguments Ryle flirts with behaviourist concepts, he is correct to point out that a dichotomy between internal and external actions is unhealthy in philosophical discourse.

This author builds on these arguments to contend that relationalism is suitable for the educational paradigm because it avoids such philosophical pitfalls. Relationalism accounts for a view of intellectualism which embraces the notion that meaningful action is a joint property between the person and the practice with which he engages. When Sarah reads, her ability to read – which is, of course, an intellectual activity – with purpose, poise and understanding of context is not a feature of some interior act of intelligence. Rather, it is an act which is as much a feature of what she reads as it is of how she reads it.

The intellectualist legend – which is, by definition, an intrinsic model of ascription – accounts only for Sarah's part of the picture, and neglects the significance of the manner in which she reads, and how it compares to the prevailing practice of reading. When Sarah reads, she engages with the practice and interacts with the modes and maxims of conduct which are set inside the framework of the practice.

Behaviourist approaches to this problem are no more appealing, by virtue of the fact that they give significance to the external only. Relationalism, however, accounts for an infusion between two inseparable facets of a holism between the subject and the object.

This much has been made clear already in this author's interpretation of Wittgenstein's philosophy in relation to the so-called inner/outer picture, as well as the discussion about the first- and third-persons. For the moment it should be clear why relationalism is so important for this author's discussion of educational attributes such as thinking, learning, understanding, intelligence, reading ability and so on. In keeping with Polanyi, Oakeshott, Ryle and Wittgenstein, this author makes the case that a relational paradigm is suitable to capture the unambiguous description and ascription of these attributes. Models which are committed to an intrinsic[4] paradigm are destined to fail because they cannot account for the infused relation between the measured subject and the measuring instrument.

### 9.3.5　Example of Clarification

Educational attributes, as it has been shown, are relational in nature insomuch as these attributes are a joint property of the person and the instrument used to measure the person. The example which has been given, reading ability, clearly highlights this. Ryle's discussion of intelligence also fits neatly inside the relational model. The concept of intelligence, or more specifically of a person's intelligence, is captured unambiguously as an interaction between the person and the mechanisms and devices used to test intelligence. It is only when one tries to separate the person from the device used to observe intelligence that one arrives at an ambiguity.

This point is a simple one to prove. Suppose Professor Plum has a PhD in mathematics. He is a much published author in his field, and he is well respected in the university where he is a professor of mathematics. He is, for all intents and purposes, an expert in his field. Avrum is a gifted student at GCSE mathematics, among other subjects. He has a keen interest in algebra, and enjoys reading about mathematical history. Avrum is given a test on 'GCSE mathematical concepts' by his teacher. For the sake of interest, Professor Plum also sits the test. As expected, Professor Plum scores 100 per cent in the test, but Avrum has prepared well, he excels himself, and also scores 100 per cent. Avrum's mother, hearing of this experiment, delights in telling her friends that Avrum is as good at mathematics as a university professor. Is she justified in making such a claim? Both people, after all, scored 100 per cent in the test; and the test, of course, is used to 'capture' some pre-existing internal faculty, 'mathematical ability'. At least, that is what the intrinsic theorist is committed to claiming.

But this seems absurd. Surely it is ridiculous to make the claim with any form of credibility that Avrum and Professor Plum are of equal mathematical ability. Avrum's mother, becoming a little embarrassed by her bold claims,

---

[4]　With reference to earlier work, neuroeducation and brain-based learning are two examples of intrinsic models of description and ascription.

weakens them slightly to claim that Avrum is as good at GCSE mathematics as a university professor. But she is still mistaken. For all she has done now is replace the attribute 'mathematical ability' with 'GCSE mathematical ability' and maintained everything else in her intrinsic paradigm. She is continuing to communicate ambiguously, all the while trying to protect a flawed concept of what it means to have an 'ability' in anything resembling an intellectual pursuit. She is falling for what Ryle has called *the dogma of the ghost in the machine*.

One simple swerve, and Avrum's mother can return to unambiguous communication. So she tries again, and tells all of her friends who are still willing to listen that Avrum is as good as Professor Plum *in relation to* the test that they both took in GCSE mathematical concepts. She makes no effort now to detach Avrum's mathematical ability from the test which was used to measure it. Her assertion, which is now relational, is the fullest description of the relata, it is completely unambiguous and it is detached from the flawed intellectualist legend.

One sees the profound importance of unambiguous communication in the avoidance of philosophical quandaries regarding the ascription and description of abilities, and indeed all such psychological phenomena. One might be excused for claiming that this is simply a linguistic ploy, an easy loophole to plug. However, to the intrinsic theorist, their inability to communicate unambiguously is precisely what leads them to make incoherent claims about abilities. The linguistic mistakes are not mere examples of *façons de parler*, rather, they are examples and manifestations of a commitment to a flawed model of description and ascription, made evident in grammatical and linguistic lapses. The carelessness of language is proof of the carelessness of the theory, not easily resolvable by a commitment to the same story, simply told in more careful language. That is, the ambiguity in the communication is induced *precisely because* the model is flawed, and a commitment to the same dogma is irreconcilable with unambiguous communication. Relationalism resolves these issues of ambiguity, and helps us to communicate in a manner which is coherent and sensible.

### 9.3.6 Beyond Putnam's semantic externalism: Relationalism and quantum philosophy

It is now sensible to directly connect some of these arguments on the inner/outer picture to quantum physics, in particular to its philosophical and conceptual underpinnings. Consider, first, what Faye (1991: 149–51) discusses regarding the so-called 'subject/object problem' of quantum physics:

> [The measurement conundrum in physics] is due to the impossibility of distinguishing between the observing and the observed parts ... which cannot be entirely separated. Likewise, in experimental psychology this means that attention should be paid to the specific rules and conditions under which the experiment is carried out.
>
> ...
>
> Bohr's thinking in this area can be set forth as follows: In psychology the situation encountered is patently the same as in quantum mechanics, for in both fields there is a difficulty of delimiting the objective content of what is observed inasmuch as the knowing subject influences the object known. (Faye 1991: 149)

These realizations not only serve as the connecting thread through this book; they also serve as a further demonstration of the power of the relational argument, and the connections between physics, psychology and, by this author's extension, education.

Furthermore, these remarks move beyond Putnam's semantic externalism towards the holism between the subject and the object, and between the measured entity and the measuring device. This is the attraction of the philosophy of quantum physics for psychology and for education: it gives access to a relational model of ascription and description, and it serves as a template for how these ideas can be developed.

Indeed, as Faye makes clear in his remarks, there is an 'impossibility' in separating the measured from the measuring device. This is, as he notes, as much a feature of psychology as it is of physics. When one seeks to meaningfully describe and ascribe psychological phenomena to a person, it is impossible – a strong assertion – to do so in a manner which delineates between the person and measuring device.

In the educational parallel, this means that abilities and intellectual faculties are only meaningfully ascribable and describable when one factors in the prevailing practice which pre-exists the measurement, and without which no meaningful ascription can be made without communicating ambiguously. Putnam's externalism gives over-credence to the external facts; relationalism accounts for an inseparable holism between the person and the practice. These primitive links require some further consideration, which will be the focus of Chapter 10.

# Education, Psychology and Physics

## 10.1 The fundamental attribution error

Gilbert and Malone (1991) offer an insight into the links between psychology and education in the guise of the so-called fundamental attribution error (FAE), or the correspondence bias (CB). Moreover, the connective between the disciplines, as it will be shown in the next section, stems from the primitive links of each of the respective disciplines and physics.

Despite the fact that Gilbert and Malone (1991: 22) base their discussion on some questionable psychological assumptions – such as an epistemically hidden (i.e. private) inner – there is scope to investigate these psychological considerations a little further. Attribution theories in psychology are focused on offering psychological insights into the motivations for people's behaviours. If John acts in such-and-such a manner, and does so with the understanding that things are thus-and-so inside the context which he acts, it is the job of the attribution theorist to explain how John understands things the way that he does.

Gilbert and Malone (1991: 22) paint a picture of the attribution theorist which sounds worryingly local, causal – that is, Newtonian and Cartesian to the core – and so this author seeks to demonstrate why the FAE highlights further, from a psychological perspective, there is a requirement to pull away from intrinsic *and* extrinsic models of description and ascription, and move towards an infused, entangled, relational model.

It is this author's contention that Gilbert and Malone (1991) have committed a category error in their discussions of the FAE, and so this brief discussion will be centred on correcting some of these errors.

Analogies between psychological phenomena and observable behaviour, and helium-filled balloons and the effect that wind has on the direction of the balloon (Gilbert and Malone 1991: 22–3) are entirely unhelpful, since they induce analogical connections between psychological and physical phenomena.

The connection between psychological phenomena and behavioural phenomena is in no way akin to the physical determinism and causality which governs the movement of a helium-filled balloon guided by the wind. These analogies fall prey to the first-person/third-person category error which this author outlined earlier in the book, and so a more suitable alternative must be established which adheres to these governing rules of ascription and description of psychological and behavioural phenomena.

However, before these issues can be resolved, the FAE and the CB must be defined explicitly. Attribution theories

> suggest that the psychological world is a mirror of the physical world and that these two are therefore penetrated by the same logic. Ordinary people seem to believe that others behave as they do because of the kinds of others they are and because of the kinds of situations in which their behaviours unfold; thus, when a person makes an attribution about another, she or he attempts to determine which of these factors – the other person or the other person's situation – played the more significant role in shaping the other person's behaviour. (Gilbert and Malone 1991: 23)

In other words, the ascriptions that we make in our third-person ascriptions are based on the dichotomy of evidence that we have at hand of a person's disposition and that person's situation at the time of ascription.

Broadly speaking, this model of ascription encompasses a view of an infused inner/outer relation which this author seeks to develop. Despite the seemingly tacit commitment (still) to a local causal model of description and ascription – where behaviour is 'caused' by events in the locality of the person only and where psychological and physiological phenomena are considered logically similar in kind – this view of ascription is almost reconcilable with this author's earlier writings on asymmetry and the inner/outer relation.

The FAE, then, occurs when this basic tenet of attribution is ignored, to the point where attributions and ascriptions are made on the assumption of (intrinsic?) dispositions of the person, when in fact these attributions should have been made in light of situational and contextual (extrinsic?) factors which surrounded the action or behaviour. Gilbert and Malone (1991: 23) warn, 'One should not explain with dispositions that which has already been explained by the situation.' The same authors continue: 'In scores of experiments, subjects have violated attribution theory's logical canon by concluding that an actor was predisposed to certain behaviours when, in fact, those behaviours were demanded by the situations in which they occurred' (Gilbert and Malone 1991: 23).

## 10.1.1 Examples of FAE

To highlight this point, some basic examples from everyday life make things a little clearer. Suppose one is driving on the motorway, only to be overtaken by an onrushing car which, as a consequence, cuts directly in front of one's car causing one to slam on the brakes to avoid collision. In such a situation, it seems reasonable to assert that any normal person would be at least a little angered by being forced to slow one's own car in order to avoid collision. In haste, one might even honk the horn to highlight one's displeasure at the rudeness of the driver. But this is an example of the FAE, because this exhibits an ascription of 'rudeness' to the driver in a manner which is divorced entirely of a consideration of context or situation. If one could see the situation from the other driver's perspective, one might find that the source of their seemingly rude driving was that they had just received news that a relative had taken ill, and they were needed at the family home as quickly as possible. The driver's hastiness in driving therefore was dictated by his urgency to arrive at his destination to tend his ill relative's needs.

Suppose, to take another example, that your friend at work has a propensity to suffer from depression. You have been made aware of this, and you have seen your friend when he is at a low ebb. A circular comes round in work warning that business has declined and there will be redundancies within the next two weeks. Ten days after receiving the circular you see your friend sitting at his desk in work, with an official letter, and he is clearly distressed. Knowing your friend's disposition to depression, you assume that it has returned.

This is a striking example of the FAE, when one ascribes to another a disposition (intrinsically), when the relata at hand suggest a clearer and more accurate description and ascription on the basis of the situation and the context (relationally).

One of the corollaries of the FAE is that there is an assumption that all behaviours and actions are preceded by dispositions, or by intrinsic qualities of the actor. When the child reads, his ability to read follows as the result from some antecedent psychological state, or intrinsic disposition which allows, or 'causes', him to read more or less effectively.

The relevance of the FAE in this instance is entirely clear: it shows from a psychological perspective the importance of infusing together the first- and third-persons, the inner and the outer, the subject and the object, and the measured entity and the measuring device. It shows, from an anecdotal point of view also, that there are shortcomings embedded into our intuitions about ascription and description of psychological attributes and phenomena.

The realization that there is a requirement for an infusion between the measured entity and the measuring device shows precisely why relationalism is required as the appropriate model of ascription and description of abilities such as reading ability and mathematical ability. In keeping with the previously outlined examples of reading, it should be clear how the FAE might find its way into the intuitive intrinsic models which this author has criticized. Indeed, if one ascribes reading ability to a person in a manner which ignores the context, the situation or indeed the item of prose which is used to 'capture' the reading ability (in other words, the measuring instrument), then one falls prey to the FAE. This error, it seems, is a result of a commitment to the view of things which holds to the dogma of intrinsic models of ascription and description, where the 'ability' is considered to be what attribution theorists in psychology have labelled 'dispositions'. The FAE therefore occurs in trying to ascribe a reading ability to a person, in a manner which is detached from the situational influences, and the measuring device used to measure it. Only relationalism avoids the inherent problems with this view of things, since what the attribution theorist calls 'dispositions' and 'situation factors'[1] are fused together into one indivisible whole in the unambiguous communication regarding abilities, intelligence and other such psychological phenomena.

## 10.2  Primitive links between psychology, physics and education

Gilbert and Malone (1991) cite a highly influential paper from psychology in 1931 by German-American psychologist Kurt Lewin – often cited as the father of social psychology – which explored the influences of a paradigm shift in physics at the time on psychology. Lewin (1931) cited the shift from Aristotelian physics to Galilean physics as a fundamental contribution to changes in psychological discourse. It is interesting that this recognition came at a time when Einstein's

---

[1]  The terms 'disposition' and 'situation factors' are perhaps a little unhelpful here, since they seem to maintain some kind of commitment to a dichotomy of inner and outer, as if these things are two separate 'causes' of outward behaviour: 'dispositions' being inner causes and 'situation factors' being outer causes. Nevertheless, no major conceptual quandaries result from this, provided one is clear about the terms involved.

relativity (1916) and the quantum revolution (of the 1920s) were starting to develop Galilean principles into what has become the quantum physics of today. It was perhaps too early for Lewin to realize the significance of his claims at the time on the connections between physics and psychology; but it has become eminently clear since – with contributions from Bohr, Oppenheimer and Stapp, for example – that the connection between the disciplines has always been of interest to physicists and psychologists alike.

Lewin (1931) identifies the shift from Aristotelian physics to Galilean physics, whereas this author posits that a similar paradigm shift has occurred from classical Newtonian physics to Bohrian (or more generally, quantum) physics which should be extended to psychology and to education. The Aristotelian–Galilean shift is, therefore, a precursor to the classical-quantum shift which has since followed, and which this author will develop over the next few sections.

Aristotle taught that physics was the fundamental study of the *essences* of objects. Objects behaved in a manner which could be 'explained' by their essence, or by the properties which they had. So, a heavy object was heavy because it was made of an essence which had 'gravity', and light objects were light because their essence had 'levity'. The behaviour of such objects, therefore, was explained by their essence. Heavy objects fell to the ground *because of* their gravity, and light objects rose *because of* their levity. It was a version of physics which, according to the times, was supported by the evidence at hand. It was also the most reasonable explanation of the relata being measured by the basic measuring instruments available at the time.

It is important to note that Aristotle's physics was a primitive version of intrinsic physics, which has been variously mutated, first until the time of Galilean physics as Lewin (1931) notes, and subsequently until the quantum physics revolution which in turn adapted Galilean relativist principles for a new physics. Despite the fact that Newtonian physics, for example, repudiates Aristotelian physics on face value, there seems to be no doubt that Aristotle's physics and Newton's physics follow one common trend: namely, they are both committed to an intrinsic model of explanation and description of physical phenomena.

Indeed, what Aristotle[2] posited as so-called 'elements' and 'causes', could be viewed as 'anthropomorphic' (Lewin 1931: 142) ancestors of Newton's ideas of causality in physics articulated in *Principia*. Lewin (1931: 142–3) holds that

---

[2]  Aristotle outlined these definitions of physics in his texts, *Physics* and *On the Heavens*, which are succinctly summarized in the work of Bertrand Russell, in his definitive text *History of Western Philosophy*.

Aristotle's physics viewed non-human phenomena as having essences of a similar kind (or nature) to human phenomena. Physics, therefore, for Aristotle was a means of explaining non-human behaviour in a mode which could be borrowed from the prevailing models of explanation of human phenomena at the time. He invoked 'elements' – earth, water, fire and air – to explain and assign 'causes' to events which took place in the earthly realm as it was known. Things 'beyond' the moon – which was Aristotle's unusual choice as a cut-off point – were seen to be of a different essence that is, made of some other 'element', which was fundamentally distinct from the four earthly elements, which Aristotle labelled as 'aether'. Causes were local, that is, within the close proximity of the behaviour, and were generated by one (or more) of the four causes, identified by Aristotle as material, formal, efficient or final.[3]

The salient point is, however, that this mode of explanation of the physical world and the laws which govern it remained in existence until Galileo and the introduction of his relativist model. Indeed, as Russell noted, this view of physics was 'extremely influential, and dominated science until the time of Galileo. ... The historian of philosophy, accordingly, must study them, in spite of the fact that hardly a sentence in either [*Physics* or *On the Heavens*] can be accepted in the light of modern science' (1946: 226).

In a similar manner, the view of the cause of human behaviour as well as that of psychology has been dominated by this Aristotelian model of physics; and, perhaps more worryingly, according to Lewin (1931), the commitment to Aristotle's physics from that of psychology remained steadfast even *after* Galileo. It seems that what Lewin (1931) means is that psychology has remained firm in its support of this primitive local, causal model of explanation of psychological and behavioural phenomena.

This author contends that this continued commitment is thanks to the furtherance in the interim period of Newtonian physics, which, fundamentally, despite being an altogether more sophisticated model of physics than that of Aristotle, nevertheless remains committed to the same *view* of physics and, in particular, to causality, as Aristotle's physics. As it has been shown already, both Oppenheimer and Stapp – among others – opined that psychology was stuck in the past with its refusal to move forward with the advancements in physics. Moreover, Bohr consistently made reference to psychology in his writings on physics, and also made strong assertions about the advancements in psychology

---

[3] The definitions of these terms are omitted, as they are largely irrelevant to the course of this discussion.

which ought to be in keeping with those in physics. These warnings have largely been ignored.

With reference to education, in an effort to make the discipline a 'science', there have been consistent appeals to (Newtonian) psychology to offer scientific credibility to the underpinnings of educational methods, observations and explanations. Furthermore, as this author has outlined already in great detail, this yearning for scientific credibility within education has since extended to the borrowing of ideas from neuroscience. The problem is, as shown in the lessons of the pioneers of the quantum physics revolution, the study of psychology needs to be updated, and as this author has shown, neuroscience continues to cling on to the fundamental errors which psychologists of the early twentieth century were being warned to discard.

### 10.2.1 Newton's failure: The corpuscular theory of light

Despite the efforts of Galileo, physics remained rooted in determinist, local, causal physics, in the main thanks to the efforts of Isaac Newton. Newton's physics was similar to that of Aristotle in the sense that it attempted to explain the phenomena at hand at the local level, with clear and identifiable causes and effects.

Despite the undoubted success of Newtonian physics in a great many aspects of describing physical phenomena, one finds a dramatic example of the failings of Newtonian physics in Newton's corpuscular theory of light. Furthermore, for the purposes of this discussion surrounding the extensions of the philosophy of physics to psychology and to education, one also sees the failure of Newton to depart from the intuitive trend of maintaining a commitment to the intrinsic, so-called 'local hidden variable' view of physics. This commitment, this author will show, has permeated psychology and education alike.

Developing Lewin's (1931) view of the connections between the flawed philosophies of physics and psychology from a perspective of a psychologist, it is worthwhile noting that similar views were levelled in detail of Newton's flawed corpuscle theory of light by prominent physicists Richard Feynman and Sir James Jeans, FRS. Both offer accounts of the failings of Newton's physics (Jeans 1930; Feynman 1985) which demonstrate the failings of Newtonianism in general, and by this author's extension, the failings of the Newtonian paradigm for psychological and educational purposes. Where Lewin (1931) contends that psychology is stuck in Aristotelian mode, this author contends in a similar manner that it is committed to the fundamental precepts of Newtonian physics.

The primitive connections of physics and psychology (and by extension, education) have been articulated quite clearly in Lewin (1931) and subsequently in Gilbert and Malone (1991). Moreover, and more importantly for the bounds of this book, education has fallen prey to the same failures.

### 10.2.2 Reflections on Newton's failings

So, it seems worthwhile to see one pertinent example of the failings of Newtonianism as articulated by Jeans (1930) and Feynman (1985), and to examine the analogical failings within psychology and education (and perhaps even in modern sciences such as neuroscience). This example comes in the guise of the aforementioned corpuscular theory of light according to Newton.

Newton's conundrum is captured in the following diagram:[4]

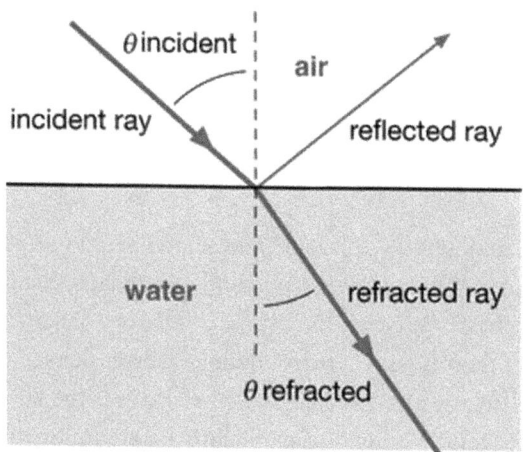

Newton's problem was as follows: on directing the light source towards the glass (or, as Newton might have discovered this phenomenon, through a river) one finds that, of 100 corpuscles (or photons) of light, 96 are permitted 'through' the glass medium, that is, they are refracted, and 4 are 'turned away' from the medium, that is, they are reflected. Newton's conundrum was, therefore: 'If all corpuscles are identical, why are some refracted, and others are reflected from the surface?'

At this most basic of stages, Feynman notes that,

---

[4]   A 'photon' and a 'corpuscle' are the same.

*Already* we are in great difficulty: how can light be *partly* reflected. Every photon ends up at A or B – how does the photon 'make up its mind' whether it should go to A or B? … This may sound like a joke, but we can't just laugh; we're going to have to explain that in terms of a theory! Partial reflection is already a deep mystery, and it was a very difficult problem for Newton. (Feynman 1985: 17–18)

It is interesting that Feynman invokes a rather apt idiom here – 'make up its mind' – with reference to how different photons will react upon being fired at the surface. This seems to be the most accurate description of our intuitions when we are at a loss to explain how things which appear the *same on the surface* begin to 'behave' in different ways. That is, when there appears to be no *external* difference, we are drawn to the intuitive conclusion that something on the *inside* is hidden from us which causes the behaviour to be different.

Now, it is clear that Feynman – nor Newton for that matter – did not really believe that photons have minds. But the idiom is an extension of the belief in what has become known in physics as a primitive version of 'local hidden variable' theory. In the absence of any visible phenomena which explain behaviour or offer a discernible cause of the behaviour (of the photon, or, under this author's extension, of the human being) one is tempted to assume that the cause and the explanation are hidden, at the local level, on the inside of the object under observation.[5]

Now, as with all theories and their supporters, Newton fought for his theory of light. He had two options open to him which allowed him to maintain a commitment to the view that he could explain the phenomena at hand objectively, unambiguously, and in a local, causal manner:

1. There may have been some strange properties in the surface that was being used for the experiment which was made up of 96 per cent 'gates', 'passages' or 'holes' which permitted easy passage of the photons through the surface, and 4 per cent made of 'spots' of an absolute reflective surface which rejected them accordingly. This argument was appealing, because it gave Newton the ability to avoid having to claim, in the first instance, that the photons weren't all identical. If one accepts that all the photons that are

---

[5] We might see this analogy develop a little further in relation to education: why is it, that if two children have been shown the formula for Pythagoras' Theorem by their teacher, and equipped with 'the same thing in mind', that they are not destined to give identical answers to all future problems on Pythagoras' Theorem? Perhaps it's because the 'thing in mind' is not guiding or causing anything at all!

emitted from the light source are identical, then surely it must be the surface with which they interact which 'causes' their behaviour to be different?

2. The other most straightforward option which seemed to be open to Newton, allowing him to maintain all other aspects of his corpuscular theory, was that the photons may have some internal mechanisms (Feynman 1985: 18) – such as 'wheels' and 'gears' – which turn in such-and-such a manner depending on the aim of the light source in relation to the surface, and thus some photons are permitted through, and other rejected from the surface depending on their internal workings upon striking the surface in question.

The rejection of each of these positions is made clear in Feynman (1985: 18–19), and the details of these rejections will be omitted.[6] The important point, however, is that Newton's attempts to resolve the outstanding issues with the problem of partly reflected light had resulted in failure. The reason for this failure was Newton's commitment to a flawed picture of the phenomena at hand. His assumption was that there was a deficiency of information available to him; if only he could 'see inside' the corpuscles, then the explanation he sought would be obvious. Objectivity was not castigated, because Newton felt that if he had the information which resided *inside each of the photons*, he could explain their strange, previously unexplained, behaviour. He was unwilling to depart from his intrinsic theory, and maintained a commitment to the view that there were local, hidden variables which, if he had access to, the problem would resolve itself.

Are not psychology and education precisely the same in that they cling to a mutated version of Newtonian philosophy? Indeed, is it not true that very often in the psychological and cognitive sciences, the staple of their experimental methodologies seems to be a search for the hidden, inner variables which might explain the outer, observable behaviour? That is, behaviour is very often sold as a poor substitute for what resides inside the individual, which, it is argued, is the source – the *true cause* – of the behaviour in question. Just as Newton was perplexed by the partly reflected light, psychologists, neuroscientists and educationalists alike yearn for an explanation of complex, and at times confusing, human behaviour. Newton's confusions led him to posit what resided inside the photons, just as the psychologist, the neuroscientist and the educationalist ponder what takes place behind the eyes when the human being behaves in

---

[6]  It is interesting to note, however, that the rejection to option (1) was due to Newton himself. See Feynman (1985: 18, footnote).

such-and-such a manner. Perhaps both Newton and his modern philosophical counterparts in the cognitive and psychological sciences would do well to remember Wittgenstein's warning that 'these things are finer spun than crude hands have any inkling of' (*RFM*, MVII, §57).

Wittgenstein opines, along a similar line of thought, that if one finds that two plants grow and bloom into two entirely different plants, *at least in appearance*, it is a mistake – although intuitive – to assume that they have germinated from two different seeds. He concludes, 'So an organism might come into being even out of something quite amorphous, as it were causelessly' (*RPP I*, §903). Furthermore, Bishop Berkeley made a similar argument in *Principles* some 200 years before Wittgenstein (Berkeley 1710, 1988, §§60–2); so too, then, with Newton's photons and, more importantly, with human behaviour.

Wittgenstein argues that the 'causal point of view' – and by this, he presumably means the concept of local causality – of science, in general often leads to some unfortunate and questionable claims (Wittgenstein 1976: 433–4). Feynman, it seems, concurs about the realistic nature of what Newtonianism failed to grasp:

> Try as we might to invent a reasonable theory … it is impossible to predict which way a given photon will go. Philosophers have said that if the same circumstances don't always produce the same results, predictions are impossible and science will collapse. Here is a circumstance … that produces different results. We cannot predict whether a given photon will arrive at A or B. All we can predict is that out of 100 photons that come down, an average of 4 will be reflected by the front of the surface. (Feynman 1985: 19)

So, what then are the consequences for science, and for scientific study? In a similar manner, what must be made of Newton's efforts to continue searching for objective results within the locality of the observation? Must we concede that science – in particular physics – is reduced to probabilities, and that determinism is lost? Feynman concludes thus:

> Does this mean that physics, a science of great exactitude, has been reduced to calculating only the *probability* of an event, and not predicting exactly what will happened? Yes. That's a retreat, but that's the way it is: Nature permits us to calculate only probabilities. Yet science has not collapsed. (ibid.)

The message is simple: descriptions of nature, not explanations, are all that is permitted in order to avoid *unambiguous communication* about the phenomena at hand. This is undoubtedly the case in physics; a science, as Feynman notes, of great exactitude.

Psychology, following on from the connections to the philosophy of physics established by Lewin (1931), is no different. The intricacies inherent in the measurement of phenomena of interest to physics, can also be extended to psychological phenomena and, moreover, to educational phenomena. Jeans (1930: 37) argues that Newton's quarrels with these problems of physics – and, in particular, his corpuscular theory of light – set down the marker for the 'abandonment of the uniformity of nature and its replacement of determinism by probabilities'.

This is precisely the remit of the quantum revolution, which will be discussed in Chapter 11 in more detail. The challenge which faced those in physics who gathered in 1911 and 1912 at the invitation of Ernst Solvay was to breathe new life into the world of physics which had stagnated under Newtonian physics for over 200 years. The view of the reality as something which was governed by determinist laws began to give way to a quantum view of probabilities and scientific *in*determinism. In a sense, one of the biggest scientific revolutions of all time was based on the view that what we all searched for – a deterministic, objective view of reality – within the sciences in particular, was, in fact, looking increasingly inaccessible; but not because of a deficiency of information or a failure of the apparatus at our disposable, rather due to the nature of what we were measuring.

Wittgenstein lambasts this in-built arrogance of science, as if all answers will become available to scientific discourse within due course: 'What a curious attitude scientists have: "We still don't know that; ... but it is knowable and it is only a matter of time before we get to know it!" As if that went without saying' (*CV*: 40). Wittgenstein seems to be dismissing the scientific agenda which undoubtedly goes hand-in-hand with the new dawn of scientism in modern times, much later than his initial time of writing. But it is also true that the sciences which maintain a commitment to the Cartesian philosophy and the Newtonian paradigm of determinism fall foul of Wittgenstein's criticism.

It is also interesting to note that Wittgenstein did not spare psychology from his criticisms:

> The confusion and barrenness of psychology is not to be explained by its being a 'young science'. ... For in psychology, there are experimental methods *and conceptual confusions*.
>
> The existence of the experimental method makes us think that we have a means of getting rid of the problems which trouble us; but problem and method pass one another by. (*PI*, PPF §371)

So, if Wittgenstein is right, we are not to assume that the answers to the unresolved questions in psychology, for example, lie further afield when, say, more sophisticated scientific methods become available with advances in technology. Rather, he is saying that when the problem is generated inside conceptual confusion, only conceptual clarification can resolve it. In the case of psychology remaining committed to the local, causal, deterministic view of things – in other words, Newtonianism – this is a conceptual issue, not an empirical one. There is a requirement, therefore, to adopt a new *conceptual framework*, not a new *methodology* or *technique*.

This, it seems, was Newton's fundamental failure: to grasp that physics was, in the main, indeterministic, that is, lacking in local causes in a great many aspects, by nature. Classical physics, of course, still has its place; when all the input variables are known, one can compute with absolute certainty what will happen next. In the game of billiards, for example, if one knows the initial velocity of the cue ball, the coefficient of friction of the cloth of the table, the angle at which the cue ball and the object ball collide, etc., then the final position of the object ball can be calculated with absolute precision. So, the point here is not that Newtonian physics is rendered useless; rather, it cannot capture what Bohr, and subsequently Honner (1987) might have called 'the description of nature'.

Wittgenstein, it seems, was pondering the very same ideas in his Cambridge lectures between 1932 and 1935, as noted by his student Alice Ambrose. After much deliberation, Wittgenstein arrives at the following conclusions about the nature of (local?) causality:

> The statement that there must be a cause shows that we have got a rule of language. … It is a question of the adopted norm of explanation. In a system of mechanics, for example, there is a system of causes, although there may be no causes in another system. … If we say we are not going to account for the changes [in measurements at different times], then we would have a system in which we have no causes. We ought not to say that there are no causes in nature, but only that we have a system in which there are no causes. *Determinism and indeterminism are properties of a system which are fixed arbitrarily.* (*WLC II*: 15–16, this author's italics)

This is an important philosophical realization. Wittgenstein is claiming here that science, like many other activities in life, is simply a language-game; that is, an activity which is governed by rules which are formed and practised in a public manner. The notion of 'causes' therefore, are part of the rules of the game which are open to being scrutinized and changed if such change is required. In a sense,

Wittgenstein is dispensing with the notion that science is a study of causes and effects; rather, that is what science has become, but this matter can easily be changed from a game which is focused on deterministic causes, and focuses more on an indeterministic system where causes are no longer of any significance. This is what Wittgenstein means when he claims, in the same excerpt:

> On hunting for the phenomenon [the cause] and not finding it, we say that it has merely not yet been found. We believe we are dealing with a natural law *a priori*, whereas we are dealing with a norm of expression that we ourselves have fixed. (ibid.)

This is precisely what the quantum revolution has achieved in modern-day science. It has changed the landscape in which science is now conducted, a shift which moves away from deterministic science, and towards indeterministic, probabilistic science.

This philosophical model is also captured in Wittgenstein's philosophies of science, where, as Cook (1994: 187–9) observes, Wittgenstein is committed to a shift from deterministic science towards indeterminism, in a move which is likely to upset our current intuitions about causality and causation; something which is also challenged in the quantum theory. Furthermore, as Cook also observes, this is not to be interpreted as Wittgenstein's denial of the right to do science; rather, it is an overhaul of the 'causal point of view' to which science seems so committed to finding in everyday life (ibid.: 188).

> This ... suggests that Wittgenstein had the following idea: science is a language game, one of whose rules is that one may always ask (and seek and answer to), 'What causes this?' But since the rules are man-made, we can always alter the rules by which we play, and in particular we could drop the rule which permits the question 'What causes this?' i.e., we could adopt an 'indeterministic system.' (Cook 1994: 188)

This is precisely what this author is suggesting: that we adopt an indeterministic system for psychology and, by extension, for education, akin to the scientific system adopted by quantum physics. This indeterminism would preclude the fundamental precepts which psychology and education alike have remained committed to from the deterministic Newtonian paradigm of physics.

It is true, also, that one of the most striking differences between Newtonianism and quantum physics is the notion of causality; in particular, the notion of 'local' causality, which will be defined in Chapter 11. Wittgenstein, it seems, was warning that the question 'What causes this?' is one which is born out of how the language-game of science is played. In this sense, therefore, the rules are open to change, and the question 'What causes this?' can drop out of the picture completely.

Cook (1994: 190) observes that Wittgenstein's view of causality is that 'we have mistaken the idea that scientists inquire after causes because *there are* causes.' That is to say, Wittgenstein seems to be opining that causes are not a fact ready to be discovered; rather, they are a feature of the game that we have constructed for ourselves, and which we actively engage in when we ask the question 'What causes this?' The fact that causes are sought after leads us to believe, wrongly according to Wittgenstein, that causes are part of nature. However, Wittgenstein rejects this notion and claims that causes are a product of the language-game called 'science'. He suggests that 'causality is at the bottom of what [physicists] do. It is really a description of the style of their investigation. Causality stands with the physicist for a style of thinking' (*WLC II*: 103–4).

Now, if Wittgenstein is right and causes drop out of the game as arbitrary, then this is why Wittgenstein concludes that it is reasonable to adopt the 'indeterministic system'. There is no doubt that physics has moved well beyond the deterministic model put forth by Aristotle and subsequently and most noticeably by Newton. With the dawn of the quantum revolution in the early 1920s, through to the modern day, there is a very clear model of science which is rooted in indeterministic concepts, and where the notions surrounding (local) causality and causation are questioned. In keeping with Wittgenstein, and indeed, with the contributions from other authors such as Lewin (1931) and Gilbert and Malone (1991), this author is contending that psychology and education must keep pace with the times. The Newtonian paradigm is no longer suitable for psychology, nor for education. Oppenheimer and Stapp warned of this many years ago. It is time their warning was heeded.

## 10.3 Interim conclusions: The within and the without give way for the between

Over the course of this chapter there has been one single prevailing trend: the 'within' of intrinsic models of explanation of behaviour, and the 'without' of so-called externalism must give way for the 'between' of relationalism.

These ideas have been developed alongside the notions of intrinsic and relational attributes, defined clearly within physics. This author has examined the inadequacies of both intrinsic psychology and Putnam's externalism, to give way to the relational paradigm of description. These ideas will be developed further in the final chapters of this book.

The natural shift in physics, brought about by the quantum revolution in the early- to mid-twentieth century, has been away from intrinsic, local causal, deterministic physics, and towards relational, non-local causal and/or non-separable, indeterministic physics. This author is contending, in keeping with Stapp, Oppenheimer, Bohr and Wittgenstein – among others – that psychology, and so too education, should follow this paradigm shift.

What would this shift bring about within psychology and education? Well, Lewin (1931) and Gilbert and Malone (1991) outline some of the ideas of how physics can influence psychology and the psychological sciences moving forward. In this chapter, this author has highlighted Gilbert and Malone's (1991) discussion of the FAE in psychology, which highlights the complications of an intrinsic view of psychology, with particular reference to how one makes attributions to others. These ideas are very closely related to previous discussions that this author has highlighted; namely, on the notions of first-person/third-person asymmetry and on the notion of an infused inner/outer relation.

Therefore, as a consequence of adopting such a paradigm shift in the psychological sciences as well as in education – away from the intrinsic models of Newtonianism (and Cartesianism), and towards relationalism – there would be a raft of complications in-built into the intrinsic models of description and ascription that would disappear. As it has been shown, Wittgenstein opines that the conceptual confusion which is inherent in intrinsic modes of psychological investigation is resolvable only with conceptual clarification. This is precisely what relationalism brings; conceptual clarity to an otherwise flawed conceptual framework within psychology, and by extension, within education.

Fundamentally, therefore, this paradigm shift would bring an end to the problematic searches for inner causes of outer behaviours. There would be a halt to viewing behaviour as a poor substitute for the inner workings of the mind or brain, as if what takes place on the inside is hidden, and we must strive to find its true content and nature. Most strikingly, perhaps, the psychological sciences, and indeed education, could begin to review the validity of the question 'What causes this?' given that 'causes', as Wittgenstein rightly points out, are a part of the scientific method, arbitrarily fixed as one of the rules of the game. If the paradigm changes towards relationalism, then so do the rules of causality and causation, and a plethora of problems disappear in an instant.

The overall consequence of this move towards relationalism, it seems, is the appreciation of the *relation* between the individual (the learner) and the practice; in other words, the relation between the measured and the measuring instrument. Intrinsic models and theories cannot account for this profoundly entangled relation. This realization is at the centre of the quantum physics scientific revolution, and it is the fundamental difference between classical and quantum physics. Any model of the psychological sciences which does not account for this feature of reality in general is, by design, conceptually flawed, and in need of updating.

These connections between the natural sciences and the psychological sciences, and by extension, education, are also not as far-fetched as they first appear. The notion which ties the underpinning philosophies of these disciplines together, as it has been shown, is the notion of *measurement* as well as the nature of the phenomena involved. It is this author's view that psychological and educational attributes such as learn, think and understand, as well as concepts such as intelligence and ability (reading, mathematics, etc.) are relational in nature in how they are measured, ascribed and described. Examples have been invoked to show that educational attributes are measureable and thus ascribable in an intelligible manner only in relation to a prevailing practice, which acts as the measuring instrument in this analogy. If one is drawn towards the prevailing scientific zeitgeist of dispensing with the importance of the practice in favour of searches of the inner confines of the mind or brain, for example, then one dispenses also with meaningful measurement and ascription of the attributes in question. In essence, relationalism accounts for the one aspect of the meaningful measurement, ascription and description of educational attributes and abilities which intrinsic theories and models can never encompass; namely, and quite trivially, human nature.

Where intrinsic (classical, Newtonian) physics deals with inert, dead matter, relational (quantum) physics deals with the active, infused relation between the measured entity and the measuring device. So too in psychology, the Newtonian paradigm of dealing in classical concepts is unsuitable for capturing the essence of active participation of the human being with the practice (the measuring instrument) which can only be truly accounted for in relational, non-Newtonian psychology. Some of these quantum concepts will be considered in Chapter 11 of this book, but for the moment the message is simple: Newtonian psychology is dead; relational psychology ought to take its place.

11

# Bohr's Philosophy of Physics and its Application to Psychology and Education

## 11.1 Introduction

In the previous two chapters (Chapters 9 and 10), there has been a detailed discussion of the so-called *relational* model of psychological attributes, in contrast with the Newtonian, intrinsic model. In this chapter, these ideas will be developed further with a particular interest in developing the primitive links between the respective philosophies of physics, psychology and education which were established previously.

This development will focus, in the main part, on the philosophical writings of quantum physicist Niels Bohr, whose ideas about physics in particular, and epistemology in general, it will be shown, are closely related to those of Wittgenstein – a connection which has been overlooked in both the literature bases on Bohr's philosophy of physics, and that of Wittgenstein's philosophy of mind and language. Indeed, in personal correspondence with Dr Peter Hacker – one of the most highly regarded Wittgensteinian scholars in the modern day – he outlined that it was entirely unlikely that Wittgenstein and Bohr would have had an awareness of each other's work, and there was indeed no documented evidence in Wittgenstein's many manuscripts (published and unpublished) to suggest that Wittgenstein had ever considered Bohr's work in physics.

Furthermore, in this author's investigation of the vast array of Bohr's writings, there are only a limited number of explicit references to Wittgenstein's works, most notably Fay and Folse (1994: 52, 94, 144, 153), Honner (1987) and Stenholm (2011), as well as a brief reference to the applicability of Wittgenstein's philosophy to quantum mechanics made explicitly in an interview with Werner Heisenberg[1].

---

[1] Full interview with David Peat, Paul Buckley and Werner Heisenberg available online at: http://www.fdavidpeat.com/interviews/heisenberg.htm.

The most useful reference for this book is found in Honner[2] (1987: 189) about the 'therapeutic' nature of both Wittgenstein's and Bohr's philosophies. Indeed, as Honner notes:

> Wittgenstein's 'therapeutic' view of philosophy [teaches us] how to prevent ourselves from getting headaches by pointing out that there are barriers to our knowing which we ought to stop banging our heads against. Philosophy can take us no further in these directions. But, and here light is cast on Bohr's position again[3], there is another side to Wittgenstein's vision. While we should remain silent about things of which we cannot speak, that is not to deny the mystical.

What Honner (1987) is highlighting here is the notion that Bohr seems to develop Wittgenstein's position (despite any explicit references to Wittgenstein's work in Bohr's writings) about what might be loosely labelled 'scientific deflationism' – or, to use a phrase already in use, *Occam's* Razor – which is steeped heavily in the important role of language and communication in how one relates scientific concepts to the everyday world in one's description of reality. Bohr, like Wittgenstein, was concerned with the restrictions which language places on how we describe the phenomena which we seek to describe. This connection between Bohr and Wittgenstein is enough to seek further development of their respective philosophies, despite the lack of interest that each seemed to have in the other's work.

What this tells us therefore, is that there are a great many interesting connections to be established at the conclusion of this book which will tie together two of the greatest philosophical thinkers of the twentieth century, in a manner which will resolve many of the outstanding issues within psychology and education alike.

It is important, however, to pause for a moment's reflection, and ponder, precisely, 'Why Bohr's philosophy of physics?' This is a pertinent question when one takes stock of the fact that Bohr's philosophy was, and indeed still is, heavily contested from within physics itself. Indeed, the most notable detractor of Bohr's ideas, perhaps rather worryingly, was none other than Einstein himself. The debate between Einstein and Bohr was, in fact, one of the greatest schisms in modern science, a division created by each of their differing views about the

---

[2] John Honner's text, *The Description of Nature*, is one of the definitive texts on Bohr's writings in both physics and philosophy. It is a fine example of a detailed and thorough exegesis of Bohr's many writings, both published and unpublished, and will act as the main text in the analysis found in this chapter of the Writing.

[3] This position will be outlined throughout this chapter.

nature of reality, and in particular, how we can communicate meaningfully about reality with the modes of observation that we have at our disposal as scientists.

Bohr's position was then, and remains now difficult to define, as one comes to realize when one sees the literature base on his writings variously describing him, using a range of labels about whether he was an anti-realist, an idealist, a transcendentalist, a Kantian or even in some cases a macro-realist (positing that only the macroscopic could be meaningfully said to exist). It seems noteworthy that Bohr himself was uncomfortable at the many efforts made by his contemporaries in physics to define him as a philosopher. Indeed, in Honner's (1987) exegesis of Bohr's many writings on physics and epistemology, he notes that many of Bohr's views were complex and difficult to formalize (23, 71–2). Moreover, there are many examples of Bohrian commentators, none more so than the eminent Abner Shimony, who often opines that Bohr's philosophical and epistemological position is almost impossible to define (Honner 1987: 23).

Bohr's placement along with some of the greatest philosophers of history by his colleague Wolfgang Pauli made Bohr noticeably uncomfortable. In a personal interview with Thomas Kuhn[4] on the day before his death, Bohr distanced himself from philosophy, claiming that 'philosophers were very odd people who really were lost.' Furthermore, one of Bohr's closest confidants and General Editor of Bohr's collected works, Léon Rosenfeld, noted in a letter to Henry Stapp that Bohr 'intensely disliked the idea of having a label stuck on him'.

So, it is precarious at the very least for this author to attempt to turn Bohr's work into a treatise on educational philosophy. The connections between physics, psychology and education are already curious enough, without invoking the work of an academic whose position in philosophical terms is notoriously difficult to define. This is the magnitude of the task which will make up the final sections of this book: to unify the ideas of Bohr – insofar as they are definable – with Wittgenstein, in the hope of building a coherent model of 'therapeutic' philosophy for education.

The appeal to Bohr's work and his view of reality and our engagement with it, one hopes, will become clearer as the chapter develops. The author will attempt to do so in a manner which does not restrict Bohr's position,

---

[4]	A great many of Bohr's writings at the time were done in personal correspondence with Pauli, Heisenberg, Einstein, Stapp, and most notably with Léon Rosenfeld, as well as many others in the quantum physics movement, particularly those from the Solvay conventions at the time. As such, much of his work remains extremely difficult to reference and record in the conventional manner. In Honner's detailed analysis of Bohr's work, many of the citations are from unpublished letters of Bohr's to his various confidants at the time.

nor in a manner which might 'stick' a label on him or his philosophy. The most important realization at this moment is, precisely, that in this author's view there are a great many connections between Bohr's thinking and that of Wittgenstein. This is where the motivation lies in adopting Bohr's philosophy; namely, that it depicts reality in a manner which is compatible with Wittgenstein's philosophy of mind and language. This, it seems was the fundamental baseline for Bohr's philosophical writings in physics: that one is constrained by language in one's description of nature. This, in truth, sounds very Wittgensteinian.

The principles of quantum physics also – and, in particular, those principles furthered in Bohr's writings (viz. complementarity, which will be defined later) – are of striking importance to this book. In almost all of Bohr's writings in physics, he makes reference to two core concepts: first, the notion which he labels 'unambiguous communication', and second, what has become known as 'complementarity', which is closely related to a corollary notion labelled in Bohr's work as 'subject/object holism'. This is what Bohr brings to this discussion in relation to psychology and education.

The fundamental concepts which Bohr furthered in physics are amenable also to education. This is the final remit of this book; to tie Bohr's work in physics to education, via his interests in psychology. The reason why this linkage is plausible, beyond what has already been argued, is simply due to the nature of the relata being measured; in physics, the quantum action; and in education, the psychological and intentional phenomena such as learning, thinking and understanding, as well as intelligence and abilities, all of which are restricted by irreducible uncertainty, and governed by rules which preclude the separation of the measuring instrument from that which is being measured if one hopes to maintain a meaningful description of nature.

## 11.2 Unambiguous communication

The basic principle of what Bohr called 'unambiguous communication' is central to his philosophy of physics, and also pivotal to Bohr's considerations on measurement and his careful efforts at a coherent description of nature. Although there are many aspects of this principle which infiltrate into Bohr's wider philosophy, the main point of interest in this case is how Bohr argued that the quantum world was accessible only through classical measuring instruments and could only be described using classical terms. This is naturally of interest to this book, because the analogy can be extended to the observation, measurement and description of psychological phenomena of interest to education.

It has been argued extensively over the course of the entirety of this book so far that language is pivotal in establishing a coherent and intelligible conceptual model inside which one conducts one's scientific inquiries. Carelessness with language leads, inevitably, to carelessness with what one might claim to be possible within the bounds of science. Thus, it has be shown, that conceptual confusion can be unravelled only with conceptual clarification, not empirical investigation. Scientific inquiry, observation and measurement, as Wittgenstein might say, is a game of linguistic exchange – a language-game – in which the rules of language and grammar are pivotal in what one claims to inquire about, observe or measure. So too Bohr was interested in how we converse in quantum physics discourse, and how such communication can remain unambiguous when the rules of language and grammar are 'borrowed', so to speak, from classical physics.

Bohr's position on the importance of language in the observation and description of nature is superbly captured by Honner (1987: 17), in an excerpt which captures the importance of all the core concepts in Bohr's epistemological, ontological and metaphysical ideas which this author will develop in line with education:

> In recent years, philosophers of science have explored the relationships between theory and observation, as well as between subject and object, producing a view of science quite different from that of the positivists. Theory is no longer seen as reducible to experience, given that the language of observation depends upon pre-existing theoretical frameworks[5] ... despite his remarks about phenomena and the crucial role of experience, Bohr is just as much concerned about the *mutuality* of the observable and observer in the whole process of observation. Through such considerations he justifies the complementary application of otherwise contradictory concepts for an exhaustive account of observations.

Honner's note here is a strong starting point for taking Bohr as a scientific philosopher whose work seems to coalesce with Wittgenstein's philosophical position with ease. The prevailing ideas are quite simple: Honner is highlighting that Bohr was intricately cautious about the nature of scientific observations and descriptions, so much so that there must be a realization within scientific

---

[5] This cyclical view of 'theory-observation-theory' has been further developed, in the main, by Thomas Kuhn, who argued that science was neither entirely *a priori*, not entirely *a posteriori*, rather an interesting infusion of both.

discussions that the language which one uses to form the description of the observation is necessarily borrowed from pre-existing conceptual schemas which are in place prior to the observation in question. Thus, in order to communicate properly within scientific discourse, in such a manner which does not induce any undue ambiguity, one must always be mindful that the language of observation and description must maintain the 'holism' (Honner 1987: 17) between the observed and the observer.

This aspect of language use is found in both Wittgenstein's and Bohr's writings. Indeed, Wittgenstein is often regarded as developing *unity of meaning over asymmetry of use* (Avramides 2001: 231) in the guise of the first-person/third-person asymmetry principle of ascription of psychological predicates. In a similar manner, Bohr is cited as developing a version of description and observation in physics which adheres to *mutual exclusion over joint completion* (Katsumori 2011: 18 and 134; Whitaker1996: 184). Both of these ideas focus on one core idea: that when one infuses the subject and the object, the inner and the outer, or the first- and the third-person, into an indivisible whole (what Bohr labelled, 'holism') one arrives at a model of ascription and description of the observable phenomena – in psychology and physics, say – which permits *unambiguous communication*, and allows for clear and coherent questions to be asked, and intelligible conceptual frameworks to be developed.

The principle of unambiguous communication is both elegant and simple. It strikes all the same chords as Wittgenstein's development of language-games, in which the rules of the game are a public possession, not a private one. Bohr's insistence on this use of public language in the discourse of physics, like Wittgenstein in philosophy and epistemology, was born from an avoidance of the pitfalls of a private language, which this author has outlined earlier in this book in line with Wittgenstein's celebrated argument on the topic. As Honner (1987: 86–7) notes, 'Because he [Bohr] was not happy with a notion of physics which was purely theoretical, he could never accept the possibility of inventing a new language for quantum physics.' Rather strangely, this was one of the main points of disagreement in Bohr's great debate with Einstein. Where Einstein was quite happy to introduce new language to physics as it was needed, Bohr saw this as a deliberate separation between the theory of physics, and the practicality of nature, which of course physicists sought to describe in the language of their study.

Bohr's argument is, therefore, that if the language used within physics displayed none of the hallmarks of physical reality, then physics did not describe anything meaningful at all. This is, in essence, a practical example of Wittgenstein's private language argument. As Honner concludes, for Bohr,

'Whatever concepts were used in quantum descriptions … must belong to the realm of classical or everyday frameworks,' outlining that Bohr stipulated this as a 'simple logical demand' (1987: 87). Indeed, as Bohr himself noted in the collected papers entitled *Atomic Physics and Human Knowledge,* 'The introduction of such unfamiliar but well-defined … abstractions in no way implies ambiguity but rather offers an instructive illustration of how the widening of the conceptual framework affords the appropriate means of eliminating subjective elements and enlarging the scope of objective description' (Bohr 2010: 70).

Bohr's principles were ingenious, because he was able to define a system of description of quantum events which adhere to this simple principle of unambiguous communication, which avoided the introduction of private – or ambiguous – terms which would make physics distinct from the reality which it sought to describe. His motive was clear: to protect objective description insofar as it could be protected; and this could only be done when the holism between the observed entity and the observing apparatus was protected, and the language used to describe them adhered to this holism. In other words, Bohr, in opposition to the so-called 'objectivists' of quantum physics – such as Max Planck for example (cf. Planck 1931) – moved to redefine objectivity in science, to avoid conceptual confusion induced by the potential of the abuse of language.

For example, as Howard notes in Faye and Folse (1994: 204), the prevalent belief within the realists in physics at the time was to assume that 'a necessary condition for scientific objectivity is the mutual independence of the scientist, as knowing subject, from the object of investigation'. However, Bohr's conceptualization of objectivity in science was to make only one simple demand: namely, that the scientist be able to communicate unambiguously about their experiments, using ordinary, classical concepts. Indeed, as Honner notes:

> [Bohr] is aware that the ordinary usage of classical descriptions only applies where the reference is unambiguous. He considers that descriptive concepts of this kind, arising as they do in the world of ordinary experience, are the only terms available to us if objectivity is to be retained in our language. (Honner 1987: 86)

Moreover, since 'communication requires some sort of sharing of experience and language' it follows that in order to communicate unambiguously 'the whole process of observer/observed interaction [must be considered], and reference must therefore be made to the conditions of the observation if the communication of what is observed is to be unambiguous and accurate' (Honner 1987: 86). If one separates the observed and the observer, therefore, one begins to communicate ambiguously, simply because one is beginning to use classical/ordinary language in a manner which induces the ambiguity. The invention of some new language, as we have seen, is off the table according to Bohr, because the description of nature requires the use of a language which is directly connected to reality, not scientifically (or otherwise) private. Therefore, Bohr's solution to the problem is to be mindful of the fact that one cannot meaningfully speak of the observed in a manner which neglects the importance of the observer or the observing apparatus. This seems like a perfectly reasonable exchange for the protection of (weak) objectivity of language.

Bohr captures this neatly, when he claims that 'the fundamental difference with respect to the analysis of phenomena in classical and in quantum physics is that in the former the interaction between the objects and the measuring instruments may be neglected or compensated for, while in the latter this interaction forms an integral part of the phenomena' (Bohr 2010: 72). As for the consequences of trying to separate the object from the measuring instrument, Bohr concludes that this would require a 'change in the experimental arrangement incompatible with the appearance of the phenomena itself' (ibid.). In other words, *it is the nature of the quantum phenomena* which makes the separation impossible, lest unambiguous communication about said phenomena be violated.

One point of note, as Howard argues, is that there is a distinction between 'sociological independence' – or observer invariance, as it may also be called – and 'metaphysical independence (Faye and Folse 1994: 205). The sociological demand is simply that as different observers carry out the observations, the results should be uniform, provided the conditions of the observations are maintained. The metaphysical (or ontological) demand – which is supported by both Planck and Einstein, among others – purports that reality is independent of the observation. Bohr, it seems, supported the need for a sociological mode of enquiry and description of nature, but made no such demands – as the realists did – about the outcomes of similar experiments (i.e. the invariance of results between the observers). The second demand – metaphysical independence – is something which Bohr undoubtedly denied, and made such disagreement clear in all of his philosophical writings in quantum physics about the true nature of reality. These philosophical positions, ardently supported by Bohr, are the presuppositions which will be adopted for this author's further writings.

## 11.3  Psychological and Educational Extensions

In psychology, and indeed in education, there is a simple analogical extension of Bohr's principle of unambiguous communication. When one makes attempts to ascribe and describe psychological phenomena such as learning, thinking, understanding, intelligence and abilities (reading, mathematics and so on) in an educational context, one must do so in adherence to unambiguous communication. As it has been shown in Chapter 10, such phenomena are categorized as relational attributes; that is, they are measured in a manner which is governed by observed/observer holism.[6] The consequence of such thinking, it has been argued, is that one can only ascribe reading ability to a person in relation to the practice of reading, and the measuring instrument (an item of prose, for example) used to 'capture' the reading ability. If one divorces the measured from the measuring instrument, one induces ambiguity in the measurement. As Bohr might argue, the statement 'John reads well' is incoherent; because it makes no reference to what John read while the attribute 'reading ability' was being observed. However, if the statement is altered slightly to, say, 'John read Dante's *Inferno* particularly well' the linguistic ambiguity is resolved, and unambiguous communication is restored.

The first statement assumes an intrinsic model of observation and measurement, and the communication about John's reading ability is ambiguous. The second statement communicates in a manner which maintains the holism between the measured (John) and the measuring instrument (the pre-existing practice of reading, and the item of prose used to do the 'measuring') and thus communicates unambiguously about John's reading ability.

This simple example, trivial as it is, highlights the profound importance of Bohr's underlying demands. It seems unimportant on face value to include a reference to the relation between the observed and the observer/apparatus, but Bohr warns that it is a necessity in order to prevent linguistic slippage. These ideas are more than the ramblings of a man with grammar *obsessive compulsive disorder*. They are far from trivialities. In the absence of these relational definitions, and in an ignorance of the principle of unambiguous communication in our scientific study, we are building our arguments on the basis of definitions which are conceptually flawed.

A failure to adhere to unambiguous communication in education, for example, leads to the view that intelligence, say, is 'within' and that the concept of 'intelligence' – some *a priori* state, presumably – can be searched for inside the mind or brain. We then begin to design 'intelligence tests' in such a manner which will capture the pre-existing inner state of intelligence, and this leads us to compare child with child, class with class, school with school and country

---

[6] A formal definition of 'holism' in quantum physics will be provided in a subsequent section.

with country. When Mike scores 67 per cent in a Geography test in which Paula scores 84 per cent, we feel assured in claiming that 'Paula is better than Mike at Geography'. When school ABC gets 85 per cent A*-C in their GCSE scores, and school XYZ gets 97 per cent A*-C, we make bold assertions about 'ABC failing' and 'XYZ achieving'. We end up with international league tables (e.g. OECD PISA) which begin to discuss a '*country's* reading ability' and a '*country's* numeracy score'. Statements are made which divorce the measured entity from the measuring instrument, and language is abused in making such questionable statements. Thus, from seemingly rather trivial beginnings, we end up in quite a muddle.

There is also the example of neuroeducation, discussed in previous parts of this book. The entire interdisciplinary programme between neuroscience and education is predicated on an inability to adhere to unambiguous communication. Indeed, the value of neuroscience to education, it seems, is to shed new light – given recent advancements in technological measuring instruments – on phenomena of interest to education, such as learning, thinking, intelligence and so on, and the explanation of such phenomena is often located in or ascribed to the brain. However, since learning, thinking and intelligence, for example, are relational in nature, a shift in the measuring instruments would lead to a change in what is measured. Neuroscience in particular, ignores this.

Therefore, the entire neuroscience programme, and indeed the entire neuroeducation collaboration is predicated on an *intrinsic* model of measurement, which posits that advancements in the measuring device will lead to a reduction of the constitutive uncertainty which is a feature of the educational phenomena in question.

The consequence is, therefore, that neuroeducation resolves nothing, and serves only to add to the ambiguous communication about educational phenomena. Tom's ability to read Dante's *Inferno* is not captured on the neural scanner, rather it is found in Tom's reading of *Inferno* with poise, awareness of context and flow. In observing Tom's reading over a period of time, and finding that he was able to read all items given to him with precision and proficiency, we might conclude, rather innocently that 'Tom is an excellent reader'. But this is not a 'numbers game', and we must always be mindful that Tom's ability to read is not a feature of him only, rather it is a feature of Tom's *interaction* with and his *participation* in the practice of reading. In the absence of the practice, Tom's reading ability cannot be meaningfully ascribed. Moreover, his ability to read, in the moments when he does not display it, is neither existent nor non-existent;

it is suspended, discontinued and returns only once he begins to read again. If we find that Tom struggles to read Milton's *Paradise Lost* with the same poise as he has previously read other texts, we need not feel compelled to review our previous claims of Tom's excellent reading ability; provided we never forget that this ability was never ascribed to him in the first place as if it were something that he carried around with him from situation to situation, inside his head, or residing in his temporal lobe.

Educational theories such as neuroeducation and brain-based learning simply get it wrong when they posit that such abilities are intrinsic to the observed individual, independent of the observing apparatus. Wittgenstein might suggest that in this particular language-game, the rules of the game have been overlooked, and there has been a conceptual blunder as a result. Bohr, similarly, might opine that the failure to account for the communitarian aspect of language in the description of the phenomena in question has led to ambiguous communication about these phenomena.

When neuroscientists, educationalists and so-called neurophilosophers claim that learning or thinking can be seen on the neural scan, they are communicating ambiguously about the relata at hand. If their error is simply a linguistic slip, then it can be easily resolved with a slight shift in position, to make such claims as 'This scan shows the activity taking place in Tom's brain while Tom is learning' or 'This scan shows the blood flow in Paula's prefrontal cortex while she decides what she would like to eat for lunch.' Unambiguous communication is restored, and the claims about the brain are no longer problematic.

However, in the cases when such claims are taken as literal – that is, when the brain is seen as the learning, thinking, intelligent organ (i.e. the agent) – then there is no way to restore unambiguous communication. Similarly, if one attempts to resolve the problems by saying that learning, thinking or intelligence are stored inside the brain, or take place inside the brain (i.e. the locus), then one will find oneself facing the same quandaries.

This type of communication about such predicates and attributes cannot adhere to the simple principle of unambiguous communication, because it does not allow for the active participation of the *human agent* with the practices of the outside world. It is essential to recognize that what is shown on the neural scanner as Tom reads, thinks or learns is not Tom's reading ability, thinking or learning; it is simply the activity which takes place in Tom's brain while he – that is, Tom, the person – engages with the practice of reading, engages in thoughtful discourse or learns, which are, as it has been argued, public possessions, not private ones.

## 11.4 Superposition

One of the fundamental ontological principles of quantum physics is the principle of superposition. In essence, the principle is concerned with the intelligibility (or otherwise) of claims of the existence of the attributes of an entity when it is measured and unmeasured. In this sense, it is closely related to, but not restricted to, Bohr's discussions about unambiguous communication and relational attributes. In fact, Bohr's discussions about these principles were his attempt to reconcile quantum physics with the classical language and descriptions which physicists already had at their disposal, in his celebrated, but heavily contested interpretation of quantum theory, *The Copenhagen Interpretation*.

The notion of superposition is, despite its simplicity on face value, a highly technical element of quantum theory. As such, these technical details will be largely omitted,[7] and the principle will be discussed only from the perspective of the conceptual framework which it provides for the purposes of this discussion.

To this end, the following excerpt from Whitaker's wonderful text on quantum physics suffices to outline the importance of superposition for this book, with primitive definitions of some technical terms to follow:

> If the wave-function does not provide a definite value for a given physical observable, that observable simply does not possess an exact value. Similarly, if the wave-function provides only a distribution of values for a physical observable ... then that observable just does not exist to a greater precision. And lastly, if the wave-function gives no information at all about an observable, that observable is totally undetermined; it just has no value. To put things another way, the statement is that no information is to be included in the theory over and above what is available in the wave-function. (Whitaker 2012: 31)

The technical term here, obviously, is the so-called wave-function in quantum mechanics. A detailed definition, once again, is not required for the bounds of this book, but in simple terms, insofar as it is possible, the wave-function is (loosely) defined thus: it is a mathematical function which, generally speaking, contains all the quantum information of a quantum system of one or more particles, when that system is considered in isolation. Whitaker (2012: 24–5) notes that 'the wave function for a particular system can tell us a considerable amount about the properties of the system, or, we should say more explicitly,

---

[7]  A detailed analysis of the principle of superposition can be found in Andrew Whitaker's *The New Quantum Age*, pp. 27–40.

what result or results we may get if we measure a particular property of the system at that particular time.'

The interesting point of note from both of Whitaker's excerpts is the notion that the wave-function in quantum mechanics seems to give some insight into what outcomes are more or less likely when certain properties of the system under observation are measured and observed. What this tells us for the purposes of this discussion is that the observable properties are not in some pre-existing fixed state, awaiting observation or measurement. Rather, it is the measurement which brings the property into a definite state of 'fixedness', rather than the measurement simply 'checking up' on some pre-existing state. The wave-function, therefore, existing in no particular state, contains *all possible states* of the quantum system, before measurement *induces* one particular state.

This is precisely what the first excerpt from Whitaker (2012: 31) means, and it gives clarity to the significance of the principle of superposition for the bounds of this book. The wave-function is a function of information about the properties of the system in question. All *unmeasured* properties are suspended in a *superposition*; that is, in no particular state, but simultaneously in all possible states that the property can take. Then, once the system is measured to examine the state of the property in question, the *measured property* of the system takes on one particular definite state. This oscillation between the indeterminate, unmeasured state and the determinate, measured state is known in quantum physics as the *wave-function collapse*; the point *in time* when a property of the measured system becomes fixed, in relation to the measurement which has been used to capture it.

The wave-function collapse and the notion of superposition have profound ontological consequences for the questions surrounding the existence of properties of a system *prior to measurement*. Indeed, in light of the principle of superposition, it becomes clear that it is unintelligible to speak of the existence of definite properties of a system prior to measurement. One cannot say that such properties do not exist, but this does not mean that they do exist in a definite state either. It is an unusual, and, indeed, counter-intuitive supposition of quantum physics that the system under observation has no definite properties nor is it in any particular state prior to measurement. This realization in quantum physics came as an ontological bombshell, not only to physics, but to science in general – namely, that measurement is the essential process which brings about the meaningful, definite existence of the particular states of a system.

These realizations are important because, as Whitaker (2012: 25) notes, it seems at first glance that the most obvious reason for finding such states during measurement seemed to be that 'if we obtained a particular value … when we performed a measurement on the system, the reason could be that the system actually had that value … even before we performed the measurement.' However, as it turns out, the prevalent philosophical position within quantum theory, thanks, in the main to Bohr and his supporters in his interpretation of quantum physics, points towards the notion that the act of measurement induces the state which the measurement itself captures. This was something – indeed, the main thing – which Einstein and Bohr disagreed over. Indeed, as Mara Beller and Arthur Fine outline in Faye and Folse (1994: 26), 'Their disagreement was over the role of measurement itself. For Einstein, measurements were prohibitive, indicating some reality already there to be measured. For Bohr, measurements became constitutive of reality.'

This interpretation of quantum physics, in one single motion, as Whitaker (2012: 31–2) opines, castigates three previously accepted philosophical positions held within classical physics, namely realism, determinism and locality. For the moment, this author will consider realism and determinism, and return to the principle of locality (or more precisely, non-locality) in Section 1.5 in this chapter.

## 1. *Realism*

Realism is the philosophical position which holds that systems have properties which are independent of measurement or observation. In support of scientific realism Einstein famously remarked in the realizations of 1935 – now famously known as the EPR (Einstein, Podolsky and Rosen) experiment – that he felt assured that the moon continued to exist even when he was not looking at it! Such a philosophical stance, in fact, led Einstein, at least in part, to reject the quantum theory. However, as Werner Heisenberg opined, 'The atoms or elementary particles themselves are not real; they form a world of potentialities or possibilities rather than one of things or facts' (1958).[8] The doctrine of realism, therefore, is quite obviously being challenged in this interpretation of quantum physics.

## 2. *Determinism*

Determinism is the notion that once we know what state a system is in 'now' (or at any point in time for that matter, since 'now' is simply an arbitrary point in

---

[8]   Page reference is unavailable.

time), we can calculate with absolute precision what will happen to it for the rest of its existence, under any circumstances. Furthermore, any two systems which are in the same state must yield the same outcomes given the same stimulus. Quantum theory was, however, to show that 'two identical systems treated identically may give *different* results' (Whitaker 2012: 32). So, determinism too had to give way within the quantum theory. Rather unusually, Einstein also rejected this caveat of the quantum theory, claiming, 'God does not play dice.' The principle of locality, discussed previously in limited detail, will be examined in more detail later in this chapter.

The salient point for the moment, however, is this: the underlying assumptions embedded in classical physics regarding the ontology of states and properties of systems are being questioned. The evidence within the experimental wing of quantum physics was, in the inception of the theory, beginning to show that the philosophical positions held within classical physics – realism, determinism and locality – were in need of review.

If realism is questioned in an educational context, in particular with reference to the nature of educational attributes and abilities such as learning, thinking, reading ability and intelligence, the consequences are clear. The realist educational position would be to contend that psychological phenomena of interest to education, such as learning, thinking and intelligence, exist independently of being measured. Moreover, a person's ability to read or do mathematics would be ontologically 'real' regardless of whether these abilities are being measured or not. In this sense, the realist must surely contend, therefore, that such things are stored within the person, for how else might they be said to exist if not inside the person? But we have seen various glaring problems with adopting this belief, none more so that the quandaries which befall intrinsic theories outlined in Chapter 10. So, to adopt superposition is to render invalid the questions which plague the realist position. We no longer ask the question: 'Are educational attributes and abilities stored in the mind or brain?' because the question no longer makes sense. Rather, we shift the focus away from the realism/idealism debate, and towards the nature of *the measurement* in education, and the jointness between the measurement, the practice and the person's participation with the practice.

Similarly, if the deterministic view of educational phenomena is questioned in line with the same questions being asked in quantum physics, then several pertinent conundrums can be resolved. Indeed, suppose, for the sake of example, two students, A and B, are considered identical in every aspect which is educationally relevant. After examination, it is found that neurologically they are identical, they have the same physiology, and, for the sake of argument, they have been exposed to the same upbringing. The students are, for all intents and purposes, identical.

Both students are now taught Pythagoras's Theorem. Their teacher tells them the formula,

$$a^2+b^2=c^2$$

Both students, after little work, are able to recite the formula. They understand what the hypotenuse is and how to identify it.

However, when it comes to the test on Friday, Student A gets the question correct, and Student B gets the question incorrect. But now we have a quandary: the educational determinist is unable to explain how this is possible. Both students have the same thing 'in mind'; the formula. But, for some reason, they are not *determined* to give the same answer in the test. Even more worryingly, another student, Student C, who has never seen the formula for Pythagoras's Theorem, is able to work out the correct answer by simply drawing the triangle with the dimensions given to him in the question. It seems that the formula here is neither necessary nor sufficient for the correct answer to be given.

What does this tell us about the determinist view in education? Quite simply that it should be abandoned. The identical students in our thought experiment, Students A and B, are not determined to give the same answer, despite the fact that they are neurologically, physiologically and otherwise identical. As it has been shown, however, by analogical extension of the principle of superposition to psychology and to education, it is perfectly possible for two identical systems to be treated identically (i.e. to be given the same teaching, for example) but to yield different results (one correct answer and one incorrect). Superposition allows for this. Whatever is considered 'identical' prior to measurement, need not continue to behave in an identical manner during measurement. This is the antithesis of determinist science, and yet it is an empirical fact in education, and indeed in quantum mechanics also.

These questions, as we have seen, are precisely those which Bohr, Stapp and Oppenheimer warned ought to be asked within psychology, and by this author's extension, within education also. The connections between measurements in physics and in psychology and, in particular, the similar nature of the phenomena in each of these disciplines, demanded that such questions be taken seriously within psychology as well as in physics.

So, it is worthwhile establishing these connections with greater verve, and the linkage is more obvious in light of two propositions from Wittgenstein, both of which have been invoked elsewhere in this book, but shall be recalled now for sake of clarity.

First recall a proposition which this author invoked in Chapter 5 to question the theoretical focus of education. Speaking about what an inner/mental object is, Wittgenstein claims:

> It's not a Something, but not a Nothing either! The conclusion was only that a Nothing would render the same service as a Something about which nothing could be said. We've only rejected the grammar which tends to force itself on us here.
>
> The paradox disappears only if we make a radical break with the idea that language always functions in one way, always serves the same purpose: to convey thoughts – which may be about houses, pains, good and evil, or whatever. (*PI*, §304)

This proposition acts as the first connective between the philosophical considerations relating to superposition in quantum physics, and the philosophy of mind, which connects directly to psychology and, by extension, to education. Wittgenstein is saying here that it is wrong to view inner/mental phenomena as the source of our behaviour; that is, as the cause of the behaviour. He claims, rather unusually, that mental phenomena are not a 'something' – that is, they are not a 'thing' about which we can meaningfully speak – but, that they are also not a 'nothing', presumably meaning that it would be wrong to say that they do not exist.

It is now timely to extend this position in line with the considerations of superposition, borrowed from quantum theory. Wittgenstein's §304 seems to be capturing the essence of superposition and serves as a philosophical definition of the principle which might be amenable to psychology and to education.

Indeed, by saying that the inner/mental phenomena are not 'something' he is saying that they have no particular form as objects *in their own right*. That is, such phenomena do not reside inside the mind or brain – whatever this might mean – and cause behaviour. But this is the first aspect of our definition of superposition; that is, the behaviour of the system (the person, in this analogy) is not to be explained independently of the measurement by a 'something' (say, an inner/mental 'cause') which pre-existed the measurement and was contained 'within' the system prior to measurement.

On the other hand, Wittgenstein also claims that this does not render valid the assertion that such phenomena are a 'nothing'. Ontologically, this assertion is also embedded into the concept of superposition. Indeed, the principle of superposition does not make claims of the non-existence of the properties or states of the system prior to measurement. Rather, it says that such properties and states are *in no particular form*. Therefore, they are not a 'nothing'. The principle of superposition, rather counter-intuitively, seems to be saying that, as physicist John Bell claimed, 'at a measurement we get a change from *and* to *or*' (Whitaker 2012: 35). Notice, this is perfectly in line with what Wittgenstein is saying about mental phenomena. We are not moving from 'nothing' to 'something' in our act of measurement; rather, we move from a superposition of *possible* states ('and') to a particular form ('or'), where the measurement induces *one particular* state.

Secondly, consider:

> We also say, 'Since yesterday I have understood this word'. 'Uninterruptedly', though? – To be sure, one can speak of an interruption of understanding.
>
> …
>
> What if one asked: When *can* you play chess? All the time? Or just while you are making a move? And the whole of chess during each move?(*PI*, §149 (a) and (b))

This proposition captures perfectly what the impact of the principle of superposition might mean in its analogical extension to psychology and to education. Indeed, Wittgenstein invokes the psychological predicate 'understanding' here, but his proposition could equally have been about learning, intelligence, reading ability or any other psychological attribute of interest to education for that matter.

What this proposition shows is the lack of continuity in ascriptions of psychological predicates. That is, what Wittgenstein seems to be saying here is that it is only intelligible to ascribe, for example, an ability to play chess to someone, when they are making a move in chess, at which point it is coherent, at least, to say 'Tom can play chess', without inducing any ambiguity.

This proposition, therefore, ties together Bohr's principle of unambiguous communication and the principle of superposition. Indeed, Wittgenstein is warning that it is unintelligible to make claims of a person's ability to play chess independent of their actually playing chess. This does not mean that they *cannot* play chess when they are not playing it; rather, there are no criteria to ascribe the ability to them in the absence of them playing it. In a rather intricate manner, then, Wittgenstein is adding further philosophical evidence to the notion that an ability – to play chess, for example – is not something which resides inside the person; but, instead, is a property which is induced by the act of measurement, say, in this instance, the act of observing the person's moves in chess over a period of time. In the absence of the practice of playing chess, and in this absence also of an ability to discriminate between correct and incorrect moves in chess, the person's ability to play chess is unintelligible.

Same goes with educational attributes such as reading ability or intelligence. This author contends that such attributes are governed by the principle of superposition for precisely the reasons captured in PI §149 cited above. Indeed, in the absence of measurement or observation, what does it mean for a person to 'have' reading ability? When one neglects the conceptual importance of the principle of superposition, one subsequently begins to start allocating the ability a location, and then one will inevitably begin to search for it wherever one located it. It is only when one makes a radical break from such incoherent thinking that one sees the profound conceptual error in claiming, as neuroscience does, for example, that such abilities reside in the brain.

In relation to studies such as neuroscience and neuroeducation, it is important to make one final remark. The principle of superposition is not to be misunderstood as applying to the brain. Indeed, it may well be that this author's extension of superposition from quantum physics to psychology and to education is misconstrued to mean that the superposition of *all possible states* is something which resides inside the mind or brain. This is not this author's contention. Rather, the 'system' (to use the phraseology from quantum physics) to which the principle of superposition applies in this analogical extension is the entire human being. Their abilities, according to the analogy, are suspended in superposition while unmeasured, and at the point of measurement, the analogical equivalent of wave-function collapse occurs, *in relation to the whole person*, at which point we can meaningfully speak of *the person* having learnt, thought, understood, read well, acted intelligently or having mathematical ability, etc.

In Wittgenstein's example, we might expand on PI §149, and note that the person can only be meaningfully said to be able to play chess, or make a particular move in chess correctly, while they are playing chess or making a correct move. As such we see an infusion of ideas beginning to make the picture entirely clear, and also how these principles of quantum physics can be readily seen in practice in psychology and in education. Indeed, we see in PI §149 the notions of relational attributes, unambiguous communication and superposition all at play. The calls from Bohr, Stapp and Oppenheimer, and their contemporaries within physics, to unify the conceptual bases of physics and psychology are looking evermore justified.

## 11.5  Subject/Object holism

The fundamental question which quantum physicists seemed most focused on in the philosophical inception of the theory was: 'What is reality?' This question seems innocent enough on first viewing, but the various answers to which it gave rise, showed that this simplest of questions opened a host of problems for physicists in establishing agreement even at the most primitive of philosophical levels. The debate, as it has been shown briefly already, seemed to centre on an even longer standing conundrum within the natural sciences in particular, and ontology and metaphysics in general, on the realist/idealist debate. In essence, the question 'What is reality?' became more about 'What does it *mean* to be real, that is, what does it mean to possess "realness?" and "Are things still "real" when they are not being observed?'

This question divided the physics community; on one side of the debate was Bohr, on the other, Einstein. This central debate was the genesis, undoubtedly, of one of the greatest debates in scientific history. On the one hand, we had Einstein, a realist, who could not accept the quantum theorist's position because it was contrary to his common sense view of reality. On the other hand, we had Bohr, in whose philosophical stance it seems wrong to label so rigidly, who contested that 'realness' was not a thing which could be meaningfully spoken about without reference to how the 'realness' is brought about in the first place, that is, without mentioning the nature of the observation or the measurement. Bohr's position, incorrectly in this author's view, is sometimes sold as his support for idealism (usually seen as Kantian or transcendental idealism; see Held (1995), Bitbol (2013)). Other authors, such as Murdoch (1990: 213–16), contest that Bohr's position is more accurately described as a 'weaker realism' with Kantian *idealist* connotations also embedded into it. Perhaps a more accurate assessment of Bohr's philosophical stance might be to say that he does not deny the existence of states prior to observation, rather he would refrain from making definitive statements about states prior to measurement or observation.

So, for example, when Einstein claimed as one of his points of contention with the quantum theory that he was sure the moon continued to exist even when he wasn't looking at it, Bohr might have retorted that he never denied that it continued to exist! Rather, a more accurate assessment of Bohr's position on this matter might be to say that it is not that the moon does not exist when one does not observe it, but one should not even concern oneself with the question of its existence in the first place, independent of its being observed. It is not the conclusion which should interest us, as opposed to the intelligibility of the question in the first place. This is not idealism; this is what Bohr called *unambiguous communication*. This is the view, as it has been shown, that if one loses sight of the restrictions of the language that one has at one's disposal, one ends up in a muddle rather rapidly.

Setting aside these simplistic arguments, there is one major point of contention which remains: namely, is existence induced by measurement, or is it independent of it? It may seem as though this question has already been answered in the discussions over the previous few sections, but it has not. Physicists have continued to argue about this for many years since the inception of quantum theory. In truth, this debate has no definitive resolution.

Let us phrase this question a little differently: 'Is it possible for the universe to exist independently from it being observed by humans?' That is, if we were not here, could the universe be *meaningfully* said to 'exist'? Einstein says, unequivocally, 'YES'! Others within the quantum movement thought differently. A close contemporary of quantum pioneers Max Born and Werner Heisenberg, Pascual Jordan claimed, 'Observations not only disturb what is measured, they produce it.' Similarly, former president of the Royal Society, Martin Rees opined, 'In the beginning there were only probabilities. The universe could only come into existence if someone observed it. It does not matter that the observers turned up several billion years later. The universe exists because we are aware of it.'

These are bold claims about the existence of the universe, and more generally about the philosophical *nature* of existence claims. Both Jordan and Rees are claiming here that the nature of existence *demands*, indeed *depends* on an observer in order to be meaningful. In truth, this is the generally accepted position within quantum mechanics, and by its leading proponents. It is also close to Bohr's philosophical position governed, perhaps more carefully than either Jordan or Rees, by unambiguous communication. Moreover, it is the main point of disagreement between Einstein and Bohr. It led Einstein to defend Special and General Relativity, with all of its classical connotations, and it led Bohr to the Copenhagen Interpretation.

The remainder of this section will be focused on examining why this author holds why Bohr was the victor of this great debate, with particular interest thereafter in showing how this most profound scientific realization can impact (analogically, at least) on the philosophical foundations of other disciplines, such as psychology and education. In the next few subsections the author will give brief descriptions of some core precepts of quantum theory which infringe on this debate, and the conclusions and analogical connections will be made thereafter.

## 11.5.1  Holism

The principle of holism, or 'subject/object holism' as it is sometimes known in quantum physics, is central to the great debate between Einstein and Bohr. Moreover, it is pivotal to the educational discussions which will follow these next few descriptive paragraphs about the analogical connections between the foundational philosophies of quantum physics and education.

Holism is a general scientific principle which posits that there is an infusion between the elements of the entire measurement situation, and a failure to recognize such infusion induces conceptual errors. The principle goes hand-in-hand with an anti-realist position in science, which posits that properties which are measured come into meaningful existence only when they are measured.

These principles are, quite obviously, related to previous discussions in this chapter and Chapter 10 of the book, in particular with reference to Bohr's relational attributes model of quantum physics, as well as the principle of superposition. In Bohr's favoured *Copenhagen Interpretation* of quantum physics, the principle of holism is pivotal. Bohr thus opined, as it has been argued, that due to the principle of holism, it is impossible to separate the observed from the observer in the act of measurement. This lack of separability between the observed subject and the observing object is therefore sometimes labelled 'subject/object holism'.

Furthermore, various theories, labelled under the term *decoherence* have spawned from the principle of holism in quantum physics, and are still heavily researched today. These theories posit, generally speaking, that meaningful (and unambiguous) communication about reality will, as a matter of necessity, invoke the measurement and the measuring instrument used during observation, precisely because it is the disturbance caused by the observation of the observed system which brings about the properties captured during measurement.

It is interesting to note that the principle of holism in quantum physics does not stop at these basic definitions. Quantum theorist David Bohm posited that the principle of holism applied not only to measurements and observations of quantum relata, but also to the entire cosmos. In line with physicists like Sir David Bates, Bohm held that the universe is often misrepresented on a macroscopic level as a collection of microscopic systems. Indeed, Bates himself often lamented the temptation of quantum theorists to trivialize quantum theory to the physics of the small. Bohm (2002: 221) argued of holism that, 'ultimately, the entire universe (with all its "particles," including those constituting human beings, their laboratories, observing instruments, etc.) has to be understood as a single undivided whole, in which analysis into separately and independently existent parts has no fundamental status.' It is unclear whether or not Bohr would have agreed with this belief.

Nevertheless, the connection between what is defined as holism in quantum physics and what has already been defined earlier in this book as inner/outer infusion in the philosophy of mind should be tentatively clear. More importantly, the implications for education are far-reaching. Indeed, with this concatenation of ideas from the respective philosophies of physics and mind, one comes to see that there is a requirement to encompass more than just the first-person, inner aspect of things when one makes efforts to talk about the phenomena of interest to education, such as learning, reading ability and intelligence. The notion of holism, when applied to educational thinking, gives rise to a model of description of such phenomena which infuses the person (the subjects, the observed), the practice (the object, the observing apparatus) and the participatory aspect of human nature (the interaction between the observed and the observing apparatus). As such, educational phenomena, such as those mentioned above, are no longer viewed as intrinsic attributes of the person, as opposed to being a feature of the holism which exists between the person and the practice.

The profound importance of this holism for education is that it brings to an end the notion that meaningful, intentional action of the human being can be viewed as something which meaningfully exists independently of the practice which brings it about. The focus, therefore, shifts away from the individual – and the theories, generally gathered under the umbrella of Cartesianism, which purport that concepts such as learning and understanding ought to be made personal and individual – and more towards the *active participation of the individual with the prevailing practice*. As such, there is a shift away from atomistic, individual-driven education, and towards communitarian, practice-driven, institution-led education, with a particular emphasis on defining concepts such as learning, understanding, particular educational abilities and intelligence, as an interaction between the person and the practice.

What precisely such a model of education might look like in practice will be discussed later in this chapter. The consequences of such thinking could impact heavily on pedagogy, instruction-style and curriculum design, among many other aspects of education in general. However, for the moment, the principle of holism lends itself nicely to this discussion for the purposes of building a conceptual framework for education which is coherent and intelligible.

## 11.6 Entanglement, non-separability, non-locality and hidden variables

When two or more systems interact with one another, in quantum physics they are thereafter referred to as an *entangled system*. The process of two systems becoming one complex system is known as *entanglement*, which in essence means that the 'states for the whole ... cannot be reduced to states for the multiple components' (Ismael and Schaffer 2014: 8). Because of entanglement, empirical evidence in quantum physics seemed to show that two spatio-temporally separated systems could behave in a manner which seemed to suggest that what happened to one had an instantaneous impact on the other. This realization is contrary to the standard held beliefs within classical physics about causality. Nevertheless, the nature of entanglement, and its empirically verified existence meant that:

> The components of a system in an entangled state behave in ways that are individually unpredictable, but jointly constrained so that it is possible to forecast with certainty how one component will behave, given information about the measurements carried out on the other(s). (ibid.)

With entanglement an empirically verified part of nature, the problem remained to explain – or at least to offer some form of possible description – of why such entangled systems were able to behave as if they were one entire system, regardless of their spatio-temporal separation. Therefore, there were three possible explanations.

### 1. Non-local causality

First, it could be that these systems were communicating/interacting in a manner which violates local causality, that is, that information might be passed from one system to another instantaneously (i.e. faster than the speed of light) in such a manner that the change in one system could bring about the change in the other.

This principle is known as the principle of *non-locality*, where causes of events need not be restricted to occurring only within a spatial boundary of the system in question. Ismael and Schaffer outline the principle thus: 'The measured results on some component(s) of an entangled system causes the other component(s) (no matter how far distant) to go into the coordinated state' (2014: 10).

Generally speaking, this idea is rejected because it violates Einstein's *Theory of Special Relativity*, which despite the fact that quantum physics is not obliged to adhere to relativity, is still a major point of contention for quantum philosophers. Einstein rejected non-local causality because he regarded it as 'spooky action at a distance,' viewing it as non-scientific and counter to common sense. Interestingly, Sir Isaac Newton also suggested that anyone who holds non-locality to be true does so contrary to basic scientific faculties. Nevertheless, the counter-intuitive nature of the principle is insufficient to preclude it as a plausible explanation of action at a distance, despite it often being sold as a form of pseudo-telepathy.

### 2. *Hidden Variables*

The second option is that there is some information contained within the system(s) which is not visible during observation, but the existence of which means that these counter-intuitive interactions occur. This is known as the theory of *hidden variables*. Einstein, it seems, was a supporter of this view. The damning evidence, however, is stacked against even Einstein, due to mathematical theorems from Gleason (1957), Bell (1964; 1981) and Kochen-Specker (1967), which proved that hidden variables theory is in fact inconsistent with quantum theory, to the point that the empirical results of quantum theory are impossible to obtain in any kind of hidden variables theory.

Tacit in Einstein's defence of this idea was his assumption of the principles of locality (outlined above) and separability (outlined in the point (3)). However, Bell (1964: 199) argued that anyone who laid claim to a hidden variables theory was forced to accept non-locality as an embedded assumption of their enquiries.[9] If Bell was correct, Einstein and his supporters were in a quandary and would have to accept that there is 'spooky action at a distance' between entangled systems; and even if Bell was incorrect, the supporters of hidden variables theory would have to make an alternative concession, and give up their assumption of classical separability. Indeed, as Faye opines in (Faye and Folse 1994: 103–18), because of Bell's inequality it is possible only to be in one of four positions: (1) a non-separability anti-realist; (2) a non-separability realist; (3) a non-locality anti-realist; or (4) a non-locality realist[10]. Einstein, because of his commitment to realism and to relativity – thus ruling out his potential acceptance of either non-locality or anti-realism – could only hope for position (2).

---

[9]  See also the citation on Ismael and Schaffer (2014: 10–1).
[10]  It is worth a note that being in favour of (1) does not preclude also being a supporter of non-locality as well as non-separability. Faye, however, does not consider this.

However, Einstein could never support the principle of non-locality[11] since it violated his ideas on relativity, so he would have to abandon the principle of separability, which he had previously taken as an assumption in the EPR experiment. This, as it will be shown, was equally problematic, due to the definition of 'physical reality' which the EPR experiment demanded.

### 3. Non-separability

The third and final option to explain this counter-intuitive phenomenon is that the two systems are not in fact, despite their spatial separation, divisible into two *separately analysable* systems. This is the principle of *non-separability*. The principle of non-separability is contrary to the classical principle of separability, which posits that 'the unmeasured particle has some reality' (Beller and Fine in Faye and Folse 1994: 23).

What this means, essentially, is that two (or more) spatio-temporally separated systems must be said to have independent reality from one another. That is, system A and system B, entangled or otherwise, which are a far distance apart (far enough that no communication can travel between them instantaneously, say) are such that the unmeasured reality of A is independent of that of B. Conversely, separability also posits that of the composite system AB, the properties of A and B are individually well-defined,[12] and are such that the properties of their composite AB are simply the composite of the properties of A and of B *individually*.

Therefore, the principle of *non*-separability, in contrast with separability, asserts that two spatio-temporally separated systems A and B, once entangled into a composite system AB, no longer have independent (unmeasured) reality. Thus, the measurement of one of the composite parts of the system, A, brings the other part, B, into reality simultaneously. Consequently, a measurement on A, as part of the AB system, brings about a simultaneous measurement on B. One might say that, once entangled, non-separability posits that it is no longer possible to conceive of the independent existence of either A or B, rather of the indivisible existence of the composite AB. The 'realness' of A now is inexorably linked to the 'realness' of B, and vice versa.

---

[11] It is interesting also that Bohr was not a supporter of non-locality either. It was in fact one of the few areas of agreement between Bohr and Einstein in their great debate. See, Mara Beller's comments in Faye and Folse (1994: 26).

[12] The term 'well-defined' is simply a mathematical way of saying that the properties which are mentioned have meaning and are intelligible.

The principle of non-separability is a corollary to holism, outlined earlier. According to Ismael and Schaffer (2014: 1, original emphasis) the definition of the principle is as follows:

> Quantum mechanics seems to portray nature as *nonseparable*. Roughly speaking, this means that quantum mechanics seems to allow two entities – call them Alice and Bob – to be in separate places, while being in states that cannot be fully specified without reference to each other. Alice herself thus seems incomplete (and likewise Bob), not an independent building block of reality, but perhaps at best a fragment of the more complete composite Alice-Bob system (and ultimately a fragment of the whole interconnected universe).

Furthermore, non-separability is crucial to the many interpretations of quantum philosophy, in accounting for the strange interaction between two spatio-temporally separated, but entangled systems. Moreover, it is one of the defining differences between the philosophical principles underpinning classical physics, and those underpinning quantum physics.

## 11.7  EPR and the definition of 'Physical Reality'

In remaining steadfast in their support of the intrinsic, hidden variables theory of physics, EPR offered the following definition of *physical reality* in their famous 1935 paper entitled 'Can quantum-mechanical description of physical reality be considered complete?':

> If, without in any way disturbing a system, we can predict with certainty (i.e. with probability equal to unity) the value of a physical quantity, then there exists an element of physical reality corresponding to this physical quantity. (EPR 1935: 777)

They also claim that this condition, although not necessary for physical reality, is sufficient for any interpretation of physics, both in classical and quantum discourse (ibid.: 778). The consequence of accepting such a definition of *physical reality* is to accept a realist view of nature. That is, the EPR criterion of physical reality essentially posits that one can only claim that a property of an object is physically 'real' if it can be proven to exist *independently of measurement or observation*. The only way this can be done is to rule out the possibility of its existence being created by observation; in other words, to accumulate a mass of evidence to show that one has predicated with absolute certainty the 'state' of the measured property, before the measurement has actually taken place. EPR conclude that if the state is not created by the act of observation or measurement, then it follows that such a state must have pre-existed the measurement or observation.

EPR claim that this criterion is generally accepted (1935: 778), and also that although it might be questioned due to its insufficient restrictiveness, any alternative criterion could not differ greatly from the criterion that they give (1935: 780). This assertion is questionable, in the main due to the fact that it tacitly assumes separability to be an *ideal of natural order*, which is in fact, contrary to EPR claims, rejected in quantum theory. Nevertheless, EPR then spend the remainder of their writing space showing that the wave-function – purported by quantum theorists to contain all the information about a system and its behaviour – to be incomplete. In other words, EPR claim to show that there is some information in the system under observation which the wave-function of the system does not contain.

They do this by claiming that, due to entanglement, considering an entangled system AB, when one composite part of the system, A, in location p(1) is measured, this act of measurement allows us to predict with certainty what state the other part of the system, B, in location p(2) will be in when it is measured. The conclusion is, therefore, that there exists two distinct, spatio-temporally separated systems, A and B, both of which must satisfy the criterion of physical reality (according to EPR, outlined above), despite the fact that only one of the systems is being measured. Therefore, if the measurement on system A appears to affect the state of system B, this must mean that the system B already satisfied the criterion of physical reality *prior* to it being measured as a system in its own right. This realization, according to EPR, proves that the quantum mechanical description of nature (in this case, represented by the simple systems A, B and AB) is incomplete; that is, it could not contain all the necessary information to describe nature fully.

Jan Faye offers a neat critique of EPR's conceptual mistake in their experiment in Faye and Folse (1994: 102). The intricate details of this critique are largely irrelevant for this book, but the salient point is this: EPR started with a criterion for physical reality – that is, a definition of what it means to be real, or to possess 'realness' – which does not extend to quantum theory. The essence of the error in the definition is that it tacitly (or otherwise) assumes the principle of classical separability with regard to the composite system AB. The assumption is that the entangled system AB can be separated and the individual parts of the system, A and B, can be measured in a manner which treats A and B as separately analysable.

However, contrary to EPR's tacit assumption, quantum theory posits that once A and B have become entangled into the composite system AB, the properties of A and B are now treated as *non*-separable; that is, where the state of A is now inexorably linked to the state of B, and vice versa. In other words it is not possible to measure A without 'disturbing' B, because AB is a non-separable system, whose properties are not necessarily a simple concatenation of those from A and those from B *individually*. Therefore, it is not that B is in some pre-existing state prior to the measurement of A; rather, when a measurement is carried out on A, the measurement is in fact carried out on the *system* AB, at the spatio-temporal point in which A resides, and from which B is far-away distinct. B, therefore, takes on a state which is *dependent* on the measurement on the system AB, at the point in space-time where-when A is measured. There are no 'hidden variables,' and there need not be any information about the system AB which is not contained in the wave-function of the system. EPR's conclusion, therefore, that quantum theory is incomplete, is predicated on a flawed definition of physical reality.

## 11.8  Complementarity

The principle of complementarity, while being Bohr's single greatest contribution to quantum theory, is variously disputed and poorly defined within the quantum physics literature. It is a principle which Bohr appealed to in order to resolve some of the paradoxes of quantum theory, such as wave–particle duality. Furthermore, the principle of complementarity was Bohr's solution to finding a bridge between objective classical physics and quantum physics.

Generally speaking, complementarity posits that the strict definition and use of concepts in quantum physics is governed by the situation in which one is observing the concept and that whatever way one records such observations and communicates about them must be in the language of classical physics. Furthermore, because the strict definition of such concepts is context dependent, it is conceivable that concepts can take on mutually exclusive forms in distinct observations. As such, Whitaker (1996: 184) opines that complementarity can be summarized as the 'mutual exclusion but joint completion' of concepts from one observational context to another.

One sees the profoundness of this principle in the aforementioned example of wave–particle duality; a paradox which troubles physicists even in the modern day. For Bohr, the problem disappears when one realizes that the complementary nature of the classical concepts 'wave' and 'particle' are both required in order to allow for the 'joint completion', full description of nature which Whitaker mentions. However, it is not conceivable that a system can portray both wave and particle properties *simultaneously*, so when one jumps from one observational context to another, one is bound by the 'mutual exclusivity' of wave and particle characteristics, which will manifest themselves *depending* on the measurement context.

Therefore, is should be clear that the principle of complementarity encompasses much of what has already been attributed to Bohr in this chapter, in particular in the discussions about his adoption of unambiguous communication, and observer–observed holism. Perhaps more importantly, however, complementarity allows pre-measurement superposition of states, and post-measurement wave-function collapse, without inducing any unwanted inconsistency in quantum physics and the efforts to apply the theory to describe nature unambiguously.

## 11.9 Some philosophical conclusions

The philosophical consequences of these principles of quantum theory are profound in how one views reality, and, in particular, how one might set about making efforts to describe nature. From the perspective of metaphysics and ontology, the philosophical perspectives offered in quantum theory strike on pertinent chords as to what can meaningfully be said to exist, and the role which measurement plays in ascribing properties to systems under observation.

The many interpretations which exist of quantum physics show clearly that there is no widespread agreement about how the ontological problems which exist as a part of nature can be resolved. It seems that depending on one's philosophical ilk, one is free to choose which position one wants to align oneself with, within reason, and justify it thereafter. As it has been argued, the positions open to us are: (1) non-separability anti-realism, (2) non-separability realism, (3) non-locality ant-realism or (4) non-locality realism.

It is not within the remit or scope of this book to make any efforts to resolve these issues. However, by adopting the principles of unambiguous communication, holism, entanglement, non-separability and complementarity, this author has thus far maintained a broad tacit acceptance of Bohr's (and

Heisenberg's) preferred position in quantum theory, known as *the Copenhagen Interpretation*.

Despite the complexity of this interpretation of quantum physics, the main appeal of it for the bounds of this book, besides its philosophical consistency, is its compatibility with the ordinary language philosophy of Wittgenstein. Everything so far considered, therefore, seems to point to the fact that Bohr's *Copenhagen Interpretation* of quantum physics is the most philosophically amenable interpretation of the description of nature for the philosophical extensions which this author wishes to make with regard to the description of the psychological phenomena of interest to education. Indeed, as this author has argued already, the constitutive uncertain nature of psychological phenomena is akin to the uncertain nature of the atomic phenomena of interest to quantum theory. Consequently, the philosophical extensions which will make up the next chapter (Chapter 12) are at least plausible.

Part Five

# The Wittgenstein-Bohr Model of Education

# A New Educational Philosophy Based on Bohr's Interpretation of Quantum Physics

## 12.1 Introduction

In this chapter, the author will make clear what the implications of such philosophical concepts, borrowed from the philosophy of physics, might have on educational discourse. Furthermore, the author's position will be made clear now, with regard to which of the philosophical positions will be adopted and indeed adapted to form the basis of a new (Wittgensteinian–Bohrian) philosophy for education, to be established fully in Chapter 13.

## 12.2 Anti-realist education

Recall that anti-realism posits that the properties of a system can only be meaningfully spoken about once an observation/ measurement has been carried out on the system. As it has been shown, this does not render valid the assertion that such properties *do not* exist prior to measurement (cf. idealism), rather they cannot be meaningfully ascribed to a system in the absence of measurement.

So, what might the implications be for education of such thinking? That is, what views does an anti-realist educationalist hold? Well, by adopting this principle and extending it from the philosophy of physics to education, one is contending that educational attributes are meaningfully ascribed only with reference to the act of measurement. It is clear that such thinking will have profound implications for many aspects of education, in particular, concepts such as assessment, teaching and learning, and on a more theoretical note, pedagogy and teaching methods.

The anti-realist educationalist will never divorce the ascription of concepts such as intelligence and understanding, for example, from the test (the measurement) which is used to measure such attributes. One can imagine, therefore, the significant impact such anti-realist thinking would have on testing and assessment, in particular on national and international comparative tests such as OECD PISA; tests which seemingly try to ascribe ability to children independent of the test, and whose creators are the most vocal detractors of so-called 'teaching-to-the-test' methods of instruction.

Moreover, with regard to teaching methods and pedagogy, there is now greater scope to offer conceptual support to the more teacher- and institution-centred and traditional forms of teaching, with one eye always being kept on the test which will be used as the measuring instrument at the end of the learning process. Indeed, those who argue that so-called teaching-to-the-test is a lesser form of teaching in comparison to more liberal teaching methods which focus on skill development and learner individuality and creativity, will find themselves in an unjustifiable position. That is, if the test is an indispensable facet of the meaningful ascription of educational attributes and abilities, it follows that teaching towards the test, of some description, is the *only* type of teaching which makes meaningful sense. Anyone who posits that the development of educational attributes can be meaningfully achieved in the absence of some form of acknowledgement of the test, therefore, is no longer safe in making such assertions.

Finally, from a more theoretical view, anti-realism castigates a wide range of educational theories – both old and new – which posit that educational attributes are a feature of the learner *only*, so-called monistic learning theories. Of interest to this book, one such theory is undoubtedly found in the neuroeducation, as well as in the brain-based learning theories in general. Such theories, in fact, being founded in an intrinsic attributes model are fundamentally realist in nature. Indeed, the neuroeducationalist or the brain-based learning theorist posits, by the very nature of their investigations, that a deeper understanding of the brain will lead to a deeper understanding of educational attributes and abilities. Thus, such theorists have two options open to them to justify why they conduct such investigations: Either (1) they adopt a realist position, and suppose that such attributes and abilities pre-exist the measurement and that an observation of the brain will enlighten the observer about some previous uncertainty regarding the brain being either the locus or the agent of such attributes and abilities; or (2) they adopt an anti-realist position, and suppose that such attributes and abilities do not pre-exist the measurement, in which case it is the neural scan, for example, which is the act of measurement which induces the state of learning, thinking, intelligence or understanding, for example.

Position (1) is rejected on the basis of previous arguments made in this book, namely the mereological fallacy and the asymmetry category error. Position (2), being anti-realist, is more plausible and is in keeping with most of what has been said in this chapter so far, and requires further investigation. If position (2) is adopted, the consequence is that one can no longer posit that the neural scan (the act of measurement) and the attribute it measures are independent of one another. What might this mean? Well, it would mean an entire redefinition of the concepts of interest to education. Learning would no longer be ascribable on the basis of behavioural criteria; rather, it would be something which is shown by a neural scan. To test if someone can read, one would have to redefine the current behavioural-based and behavioural-dependent definition of 'reading ability' where the measuring instrument is, for example, and item of prose and the existing practice of reading, and shift towards a definition of reading ability which encompasses an interpretation of a neural scan. One could be said to read, therefore, only when one's neural scan said so.

Such a redefinition of educational attributes is not prohibited. To be clear, however, the educational landscape would dramatically change. Every child would have to be fitted with a neural scanner, and when the teacher wishes to test their ability to solve 'Type A' mathematical problems, they would have to be trained to interpret the neural scanner and deduce from it the meaningful, accepted, common interpretation of what it means, neurologically speaking, to be able to solve 'Type A' mathematical problems.[1] Reading a book with proficiency and elegance would no longer be the criterion required for ascribing reading ability; the measuring instrument would shift from a community-based practice, and towards a neural-based picture of blood flow and colourful diagrams. Our relational and anti-realist position remaining intact, this is the *only* option for the anti-realist neuroeducationalist and brain-based learning theorist: to take a neural scan in place of observable, communitarian behaviour as the pre-set measuring device, and acknowledge that human behaviour is no longer of *any* significance in the meaningful ascription of educational attributes and abilities.

To take this one stage further, for the sake of clarity, one must ask oneself: What are the implications of such redefinitions for everyday (educational) life? Quite simply, education would become unrecognizable. Children would no longer read, do mathematics or learn a second language at school; that is, not in the way in which we currently understand those things. The anti-realist who has

---

[1]  See Harré and Tissaw (2005: 94) for Wittgenstein's *modus tollens* argument against this possibility.

redefined reading as something which shows up on a neural scan has changed the concept of what *it means* to read. In fact, to take this point to the extreme, classrooms would now be filled with children who no longer have to speak! Rather they ruminate over a problem in their mathematics class, and when certain colours are found on the neural scan, this means that they are solving problems in the accepted manner. The template for solving problems correctly now will be a neural scan, and only when the child's neural scan matches the template neural scan can they be said to be reading, solving problems, or acting with intelligence and understanding. As outlandish as this sounds – and no one, of course, is proposing this – the reality is that if neuroscience seeks to redefine educational concepts, this is ultimately what would have to happen.

This sounds all a little far-fetched, more fitting for the realm of science fiction than for considerations in meaningful scientific discourse. The reality is that such a position ought to be discarded on the basis of the fact that education is a weak science lacking in strict objectivity, and this is by-and-large due to the complexities of the human nature. When human beings interact with one another, as they invariably do in education, there is an aspect of this human interaction which is unamenable to scientific explanation or description. The neural redefinition of educational attributes and abilities does not allow for this human interaction, and as such, restricts the model of explanation and description too far; so much so that none of the human remnants remain in such an explanatory or descriptive model. The anthropological and sociological nature of the human being would thus be lost when one moves education away from the social sciences and towards the brain sciences.

Furthermore, such a paradigm shift within education ought hardly to be encouraged, since, apart from the obvious, there would still remain an interesting logical gap which the educational neuroscientists would have to 'plug' before anything meaningful could ever be said in such an educational landscape: namely, that the *correlations* which neural scanners show are not (and indeed never can be) considered equivalent to *causations*. In other words, even in the world where neural scanners are used to identify the neural processes which take place when a person, for example, reads, these processes are, by their very definition, simply *correlated to* the person reading aloud. It is a logical fallacy to conclude that the reading is caused by whatever is taking place in the brain while the person reads. Such a logical flaw comes when one attempts to flow up the causal stream (from the reading, towards the brain activity, resulting in the claim that the brain activity has downwardly 'caused' the reading which it depended on for its identification).

Therefore, the challenge for the educational neuroscientist would become to develop a new type of neural scanner which could show *causations* rather than correlations; a feat only achievable if it were conceivable that such a machine could be developed to identify the neurophysiological processes *independently of* the person having to read aloud when prompted.

But, at this stage, we are in something of a muddle, for even in the case where there are underlying brain processes which could be identified independently of reading prose aloud, we remain at a loss to explain how having such brain processes equip us to do anything at all worth note. That is, as Harré and Tissaw (2005: 87) remark, 'In order for something in mind (or brain) to be foundation for a skill, we must know how that something is to be used.' That is to say, we have explained nothing by saying reading ability, for example, is a brain process; since if it is a brain process which does not accompany a *use* (i.e. a manifestation), then it cannot be meaningfully said to be an ability of any kind. In relation to the causation problem outlined above, Harré and Tissaw (2005: 88) also opine, 'An ability is not a special state of mind. An ability is displayed in public performance.' Therefore, although it is clear that the brain is active while a person reads, and that the brain is necessary for such reading activity, it does not follow that concepts like 'reading ability' (or any other ability of interest to education for that matter) can be captured on the neural scanner. We are implored to remember Professor Raymond Tallis's warning to neuroscience (in direct relation to their standard practice) that 'correlations are not causes'.[2]

The overall conclusion is, therefore, that the anti-realist position in education which this author posits, leaves open only one plausible option: namely, that we ought to focus more on the *interaction* between the human being and the traditions and practices which are public possessions, and move away from theories such as neuroeducation and brain-based learning which posit that an examination of the brain will shed new and meaningful light on some previously undiscovered scientific marvel. Moreover, holding to the anti-realist mantra borrowed from the philosophy of physics, this author contends that neuroeducation and brain-based learning are insufficient models of description of educational attributes and abilities; attributes and abilities which are, by definition, relational in nature, and therefore requiring reference to the measuring instrument and process used to capture them. The natural choice for the measuring processes in an educational sense must in some way account for

---

2  Cited in a lecture, accessed from https://www.youtube.com/watch?v=ku-GmndXDXo

the pre-existing, communitarian established traditions, practices, institutions and forms of life; all public possessions, not private ones.

## 12.3  Indeterministic education

As it has been shown, one of the fundamental differences between classical physics and quantum physics is the philosophical shift from determinism (classical) to indeterminism (quantum). Classical physics remains perfectly sufficient for calculating, say the velocity with which a billiard ball will strike the cushion in a game of billiards, provided we have the necessary information in order to carry out the calculation, such as the velocity of the cue ball on impact with the object ball and the coefficient of friction of the cloth on the table. The quantum world, however, is a little different. Quantum indeterminacy is constitutive of the atomic phenomena in which quantum physics is interested. The nature of quantum predicates dictates that such indeterminacy cannot be reduced *prior to measurement*. Thus, the best that we can hope for is a collection of possible outcomes, each with a prescribed probability (known as a state of superposition) dictated by the information which is contained in the wavefunction of the quantum system. The important thing to realize is that *the act of measurement* is what makes the indeterminate become determinate.

This description is equally applicable to the attributes and abilities of interest to education. Indeed, transcribing these philosophical ideas onto educational indeterminism would resolve a host of education quandaries in an instant. Most significantly, perhaps, it would allow educationalists to oscillate between the indeterminate *pre*-measured human being and the determinate *post*-measurement human being. In a pre-measured state, the human being would not be said to have any definite educational attributes or abilities. Such attributes and abilities, being relational in nature, allow the indeterminist educationalist to contend that it is the act of measurement which is instrumental in meaningfully ascribing these attributes and abilities to the human being, *post*-measurement.

Moreover, the indeterminist is not committed to ascribing such attributes and abilities to the person in the absence of measurement, and so they avoid the philosophical complications with having to answer such questions as: 'Where do these attributes and abilities reside within the person?'

Furthermore, when one comes to embrace the notion that pre-measured states of educational attributes and abilities are indeterminate, one can begin to let go of the flawed causal set-up of the inner/outer picture discussed earlier in relation to Wittgenstein's writings on the matter. Indeed, as it would turn out, supposing educational indeterminism is adopted, it would no longer be the case that behaviour would be viewed as a poor substitute for something which goes on behind the eyes, rather one would come to embrace that in a pre-measured state, *the entire human being* has indeterminate educational attributes and abilities, and in a post-measurement state, *the entire human being* has attributes and abilities which take on definite values and manifestations. The 'hidden' is no longer seen as a cause which resides on the inside, as opposed to simply being a feature of the entire unmeasured system. There is no dichotomy of 'inner' and 'outer', standing apart in a local causal relation, rather a distinction between that which has not been measured – and is therefore indeterminate – and that which is measured – and is therefore determinate. A cloud of conceptual confusion disappears in an instant.

## 12.4 Holism in education

The concept of holism – or subject/object holism – is perhaps the most basic extension to establish between the philosophy of quantum physics and that of education. As it has been shown already, holism posits that only once the observed subject and the observing object (the measuring instrument, together with the observer) are brought into interaction with one another can one meaningfully ascribe properties to the observed object in question. So, after this interaction, which is induced by the measurement in question, the subject and the object form an indivisible whole, to the extent that the ascription of properties to the observed system is not made with reference to the single observed subject only, rather to the entire subject/object system. Embedded into the principle of holism is the concept of unambiguous communication, both of which taken together form part of the description which is captured in the relational attributes model outlined in Chapter 11.

The philosophical analogy for education is a simple one. In education, the observed subject is naturally the (individual) learner, with the observing/measuring instrument being a combination of the educational test which is being used, and the practice/tradition of the subject matter which is being tested. What the principle of holism means, therefore, for education is that all educational attributes are ascribed to the person with reference only to the test/practice which is being used to measure. This redefines concepts like 'reading ability' and 'intelligence', which will no longer be seen as a property of the individual learner, rather a joint property of the interaction between the learner and the test/practice. In this way, it is clear that holism is the underlying principle which supports the thesis of relationalism in education.

On a primitive level, this means that practices and traditions, such as mathematics and literature, for example, take precedence over the atomistic, individual view of child-centred education. Claims of fitting the school to the child rather than the child to the school no longer carry any substance, since it is the practices and traditions – as well as the communities and institutions which are established in a school context – which are of paramount importance. Education, therefore, should be viewed as the inculcation of the learner into a form of life, a manner of working which is communitarian, anthropological and sociological in nature. The separation of the learner and the test/practice in trying to ascribe educational attributes and abilities to the learner (as an individual) will lead only to conceptual confusion. If one removes the practices and traditions from the holism between the subject and the object, one arrives only at nonsense.

On a wider view, and of particular interest for this book, it is clear that theories such as brain-based learning and neuroeducation are predicated on a misconceived view of educational attributes and abilities. When the brain-based learning theorist posits that the brain can learn, he is mistaken, because learning is now defined as the holistic relation and interaction between the learner and the practice of whatever is being learnt. When the neuroeducationalist claims that the learner's mathematical ability can be seen on the neural scan, he is in a muddle, since mathematical ability is a joint property of the learner and the practice and tradition of mathematics, which cannot be captured on a neural scan.

In fact, in light of the principle of educational holism, the appeal to brain science at all is rendered practically useless, since the meaningful ascription of educational abilities and attributes *in no way* depends on the neurophysiology or neurobiology of the learner; it is, in fact, a joint property of the learner, and his participation in the practices and traditions of life. The ascription of *correct* – a keyword – conduct, a precursor to the ascription of educational attributes and abilities, is not found on the neural scan, nor on some new knowledge of the brain and its structure; rather, it is found in the learner's active participation with a form of life, which is adopted by a community of like-minded masters, who cultivate the practices and traditions which *define* such correct conduct. Only when the learner *acts in agreement* with these practices and traditions, and when his judgements are the same as those of his master, is he meaningfully said to have such-and-such an ability or attribute. As Kripke (1982: 106) opines[3] in his detailed exegesis of Wittgenstein's work on rules and private language:

> Yet how does agreement operate? ... How do we judge of someone else that he has mastered [a concept]? Our judgement, as usual, stems from the fact that he agrees with us in enough particular cases (and that, even if he disagrees, we are operating with a common procedure). ... We do not compare his mind with some suprasensible, infinite reality. ... Rather we check his observable responses to particular ... problems to see if his responses agree with ours.

This, however, it should be noted, is not support for the behaviourist view of education, for as Kripke (1982: 107) remarks, in keeping with Wittgenstein's arguments found in his famous *private language argument*, 'Such [behaviourist] opinions are misguided: they are attempts to repudiate our ordinary language game.' And, as Kripke concludes, where behaviourism denies the existence or legitimacy of the 'inner' realm, Wittgenstein's discussions on the matter are altogether more careful.

Wittgenstein's only demand is that the shared judgements of like-minded masters of a practice are the benchmark for correctness, rather than some inner guiding mechanism. The precedence, as with the principle of holism, is afforded to the practices and traditions of communitarian linguistic exchange (meaning any meaningful engagement with a common language). One can become a master within a community by offering consistent displays of the mastery of the traditions and practices which the community adopts. Thus, it is these very practices and traditions which form an indispensable facet of education, and, moreover, a fundamental component of what it means to ascribe educational abilities and attributes to a person. This is, by definition, educational holism: the inseparable bond between the learner (the subject) and the traditions and practices of the community (the object) which makes the meaningful ascription of educational attributes and abilities possible.

## 12.5 Entanglement, non-separabilist education

As it has been shown, where holism applies to the inseparable infusion between the measured and the measuring device, the principle of entanglement applies

---

[3] In this citation, Kripke is referring directly to an example of how one acquires mathematical proficiency. However, the argument can be extended to any educational attribute or ability, previously mentioned in this chapter.

to the interaction between two individual systems; the notion being that once two systems interact with one another, the behaviour of the individual systems is now supervened upon by the infused behaviour of the entangled system.

Furthermore, one possible explanation for why this phenomenon occurs in quantum physics is non-separability, which holds that once two systems, say A and B, have interacted, the 'realness' of A and B is no longer intelligible, only the 'realness' of the entangled system AB is meaningful. The conclusion for the non-separabilist, therefore, is that once entangled, A and B are henceforth considered as an indivisible whole – namely, AB – the properties of which are not necessarily a simple concatenation of the individual properties of A and of B. Thus, a measurement in A – once it forms part of the system AB – will automatically induce a simultaneous change in the state of B, simply because A and B, regardless of their spatio-temporal separation, are not separably analysable.

So too for education, it is possible to draw a simple yet profound analogy. Indeed, if one assumes the roles of the two individual systems A and B are taken on by the teacher and the pupil, then the entangled system in our analogy would be the compound teacher–pupil system, entangled, perhaps – to extend the analogy – at the point of teaching/learning.

What does the principle of non-separability tell us, in the most basic sense, about the properties of this new entangled teacher–pupil system? Well, on face value, it suggests that the observation of the pupil in particular is, in fact, an observation which must account for the entangled teacher–pupil system, where the properties one then ascribes to the pupil are in fact properties of the teacher–pupil system. In other words, once the teacher has taught the pupil, and once the pupil has learnt from his teacher, they are no longer separately analysable. By extension, this would mean that the educational non-separabilist would posit that the pupil is no longer a single, individual system; rather, all of his future measureable educational actions, abilities and attributes are thus constrained (but not determined) to act in accordance with his non-separable relation with his teacher.

The principle of non-separability, therefore, seems to act on an individual level, with a particular focus on interactions between the key players in educational exchanges. This, as it will be shown in the next section, is in contrast to the philosophical concepts which govern the roles which the more communitarian aspects of education – such as traditions, practices and forms of life – play in educational interactions.

What this principle means, however, for education is a clear support for the centrality of the role of the teacher in teaching and learning. The neo-

Rousseauian notion that the teacher should be demoted to the role of mere educational facilitator is castigated under this view, with extra-philosophical clout added to the importance of viewing the teacher as a master. Indeed, if the ascription of educational attributes and abilities to a pupil is meaningful only in adherence to the principle of non-separability, then it naturally follows that the relation between the pupil and the teacher is of paramount importance. Indeed, what non-separability suggests in this instance is that the meaningful ascription of attributes and properties to a part of the system must never be done without reference to the entire system, rather than to the constituent parts of the system. Thus, one sees further support for a traditional educational paradigm in the principle of non-separability.

One final note of interest to this book is on the consequences of non-separability for the role which the cognitive sciences can play in educational discussion, in particular with regard to neuroeducation and brain-based learning theories. Non-separabilist education serves to diminish the significance of the questions which neuroeducationalists and brain-based learning theorists are prone to asking. Indeed, viewing the meaningful ascription of educational attributes and abilities as governed by the principle of non-separability renders invalid the neuroeducational and brain-based learning notion that such attributes and abilities can be explained, described and ascribed with reference only to the individual learner, via searches of the brain or otherwise. By adopting non-separability as a governing principle of educational philosophy, in particular in how one communicates about educational phenomena, one comes to realize that the cognitive science revolution in education is fundamentally restricted. It can say nothing meaningful about the ascription of educational attributes and abilities to the learner, simply because it cannot account from the non-separable nature of the teacher–pupil system.

## 12.6 Non-local education

Despite the fact that many physicists view non-separability and non-locality as mutually exclusive principles within the philosophy of physics (see Faye in Faye and Folse (1994: 103–18), for example), this author does not subscribe to such a view.

It is also true that others argue that the principles are logically equivalent, or at least – as this author posits – that non-locality is simply a principle which is embedded into the principle of non-separability.

Nevertheless, the importance of the philosophical principle known as non-locality, outlined earlier in this chapter, for educational purposes is worth investigation. Recall from the previous subsection, that educational non-separability applies to the interactions between individuals (systems) and their relations to one another. The principle of non-locality, in contrast, is perhaps more readily seen not in individual interactions between educational 'systems' – for example, teachers and pupils – rather, it applies more to the 'action at a distance', as Einstein called it, which is resultant from traditions, histories, past engagements, practices and forms of life.

Non-locality, when applied to educational philosophy analogically, posits that there exists a continuing (causal?) influence, which acts as a constraint on local behaviour, but which acts from a non-local spatio-temporal situation. In other words, there exists some other 'information' beyond that which is attributable to the information which could conceivably be within the locality of the individual learner (the system).

This notion, as it has been shown, flies contrary to the local hidden variable theory, which in an educational sense of interest to this book is found in theories like neuroeducation and brain-based learning theories, which contend that all of the facts of the learner's behaviour are found within the learner himself, that is, inside the brain, for example. What the non-localist educationalist would argue, however, is that there is non-local information which constrains ('cause' might be too strong a word here) the learner's actions from a distance.

Now, what might such non-local information look like in an educational sense? It seems that the most natural extension of this principle from quantum physics to education comes in the guise of the influential role which practices, traditions, and forms of life would play on constraining the pupil's/learner's educational and learning endeavours.

For the sake of clarity, it is worth considering the chronology of the learning process, and thus take note of where in this picture the principle of non-locality might apply. Suppose Tom, the learner in this example, is taught how to find the roots of polynomials using the remainder theorem in class on Monday. He is told that he will be tested on this topic on Friday. It is a matter of interest in education how Tom will be able to move from Monday to Friday, and be able to solve the problems on Friday with the information and judgement that he has developed in Monday's teaching. The answer, it seems, lies in the principle of non-locality, so-called spooky action at a distance.

Indeed, in this situation, the non-local description of the events which take place on Friday is able to take into account the constraints placed on Friday's behaviours by Monday's events. Recall, of course, that non-locality posits that despite the spatio-temporal separation of the learner on Monday from himself on Friday, there is some non-local constraint placed on his behaviour, acting, as it were, *at a distance*.

What this tells us therefore, in an educational sense is that on Monday, the pupil is being inculcated into a practice, a tradition and a form of life; he is being shown how to act in such-and-such circumstances, when such-and-such a problem arises. Henceforth, when he is faced with a similar type of problem to that which he has been introduced in Monday's teaching, his judgements – provided the teaching has been successful – are thereafter constrained by his previous behaviours. On Friday, therefore, Tom is brought to the cusp of determined behaviour because of this inculcation into the practices and traditions of mathematics, in particular in how to find the roots of polynomials using the remainder theorem. When Tom answers the questions on Friday, it is almost as if the teacher is invisibly present at his shoulder, acting from a distance.

In finding that Tom answered question 3 incorrectly, Tom's teacher might say, 'That's not what we learned on Monday!' What this tells us is that when Tom's ability to solve polynomials is measured using Friday's test, he is non-locally influenced by Monday's activities, and such influences are clearly non-local due to their spatio-temporal separation from Friday's events. The point in time when Tom answers a question correctly in Friday's test is non-locally influenced by his active participation with the practices and traditions of which he is now himself a part.

But it is important to note one caveat of non-locality: namely, this principle does not suggest that there is some kind of mystical 'transfer' of information from Monday (the past) to Friday (the present), by means of some magical or science-fiction time warp. Rather, the principle of non-locality adheres to a kind of discontinuity in the ascriptions of properties to systems, which in this case would be the meaningful ascription of 'ability to solve polynomials' to Tom between Monday and Friday. In other words, by adopting non-locality, one obviates the need to explain the concept of 'transfer' at all, meaning that one is no longer compelled to answer the questions which might, for example, be of interest to the neuroeducationalist and the brain-based learning theorist, whose position forces them to ask questions like 'How can Tom carry the information from Monday to Friday?' and 'Where does such information reside between Monday and Friday?' The non-locality educationalist, however, is able to note simply that Tom's inculcation into the practices and traditions of mathematics on Monday will henceforth non-locally influence his future engagements in mathematical activities of a similar kind. He does not 'carry' anything from Monday to Friday, and there is no 'transfer' of any kind from Monday to Friday. Rather, at two distinctive spatio-temporal points, Tom's behaviours are constrained (but not determined) by his initiation into a form of life.

This argument rests entirely on the realization that *correct* conduct is *a* practice, and it is practised. To act in a correct manner is not to get something *in mind*, rather it is to adopt the practices and traditions shown to one in one's training, and to discipline oneself in one's judgements to move from one context to another, in distinct moments in time, being able to identify the correct action at the correct time (which is the very definition of 'judgement'). The principle of non-local influence is a feature of the development of such judgement, and thus the ability to act correctly.

## 12.7 Complementarity in education

The final principle of quantum philosophy which can be extended analogically to education is complementarity which is often sold as the defining contribution by Bohr to the quantum debate on the nature of reality, and our tussle to describe it. The profoundness of the principle for educational purposes is in the resolution of one potential problem which might trouble the sceptic who contests that what this author has posited over the previous few paragraphs is contradictory.

Indeed, the previous arguments to extend the principles of holism, non-separability and non-locality to educational philosophy seem to be predicated on the notion that one can oscillate between two mutually exclusive 'states of affairs' when moving from the unmeasured to the measured. It has been this author's argument that it is possible to move from a state of indeterminism in the pre-measurement state, towards a state of determinism in the post-measurement state, so that, for example, a person could not be meaningfully said to have a reading ability prior to measurement, but then have a very definite reading ability ascribed to them after measurement has taken place. Moreover in the mathematics example invoked to elucidate how Tom might move from Monday's teaching to Friday's test and be able to answer questions correctly in the situations which present themselves after Monday's teaching, this author espoused the view that it is perfectly reasonable to adopt a discontinuous view of educational attributes and abilities, where one can oscillate between the 'and' of being in all possible states (pre-measurement) to the 'or', where one is in one particular state (post-measurement). As such, underpinning these arguments were also the analogical educational concepts of superposition (pre-measured 'and') and wave-function collapse (post-measured 'or').

However, the concern here is that one is in the tricky philosophical position of trying to explain how this constant oscillation between two mutually exclusive ('and' and 'or') situations can occur. The answer is, simply: complementarity, which, it should be recalled, Whitaker defines as 'mutual exclusion but joint completion' (1996: 184), and which Bell outlined as the movement from 'and' to 'or' during the act of measurement (Whitaker 2012: 35).

In his 'private'[4] moment, the pupil is *both* correct *and* incorrect in how he might solve mathematical problems, or how he reads an item of prose. However, at the point of measurement – in his 'public' moment – the pupil is *either* correct *or* incorrect in how he solves problems, reads or demonstrates any kind of behaviour of interest to education. What complementarity tells us, in an educational sense, is precisely that such a movement between two complementary states is not only intelligible and coherent, but it is, in fact, necessary to allow both states in order to attain the fullest description of nature which is possible. That is what Whitaker means by 'joint completion': the permission of two contrary but complementary states, which when taken together, and permitted only from one context to another, form a true description of the phenomenon in question. One sees this phenomenon in physics, in particular in the paradox of wave–particle duality, and so too in education in the movement from 'private' viability to 'public' correctness.

## 12.8  Looking forward…

In the final chapter of this book, one thing remains to be considered: namely, the connections between Bohr and Wittgenstein, and how each adds something to the development of a new philosophy of education in their respective philosophical discussions from within their own fields of study. The overarching aim is to resolve the long-standing educational paradox known as 'the Meno paradox' (sometimes also called 'the learning paradox'), and much of what has been written in this chapter will be invoked to that end.

---

[4]  Remember, 'private' and 'public' are inside scare quotes here, because the author is not suggesting that there are truly 'private', that is, epistemically hidden, moments; rather, 'private' here refers simply to the moments of individual reflection, when such reflection is concealed, and not shown.

# Conclusions: Wittgenstein-Bohr
# Model of Education

## 13.1  Meno's paradox

One of the most troubling discussions in educational philosophy with profound realizations for the nature of educational theory is a long-standing conundrum known as Meno's Paradox, originally articulated in the *Meno dialogues* between Plato and Socrates. Waterfield (2005: 113) outlines the paradox thus, in his account of the *Meno*:

> And how will you search for something, Socrates, when you don't know what it is at all? I mean which of the things you don't know will you take in advance and search for, when you don't know what it is? Or even if you come right up against it, how will you know it's the unknown thing you're looking for?

Socrates's rather damning conclusion in his discussion with Meno is, 'it is impossible for a man to search for either what he knows or for what he doesn't know' (Waterfield 2005: 113).

So, first, why is this paradox of learning in particular, and education in general, of interest to this book? Well, for a more detailed discussion of the matter, the reader can refer to this author's discussion of the paradox in Kitchen (2014: 62–7). However, for the moment, what Meno's Paradox outlines is the problem with viewing learning as either (1) *self*-directed inquiry or (2) accessing some pre-existing knowledge from within. Indeed, Socrates's own conclusion makes this very clear, when he claims that it is impossible for a man to be said to search for, inquire about and thus to have meaningfully learnt that which he already knows, and that it is also, and more obviously, impossible for him to search for, inquire about, and thus learn that about which he has no knowledge.

But therein lies the paradox. Indeed, with regard to any subject which we would seek to learn or understand, either one knows (partially, fully or otherwise) or one does not know the subject. However, if one knows the subject, one cannot be said to learn the subject about which one already has knowledge. Conversely, if one does not know the subject, one cannot inquire to know the subject, because one would have to be able to identify that which one previously did not know, but which one seeks to know and learn about. So, what is the resolution? Is Meno suggesting that learning is impossible? As educational philosopher Carl Bereiter notes: 'The learning paradox may be viewed as shaking the foundations of educational thought by demonstrating that the supposed role of learning (and hence education) is an illusion' (Bereiter 1985: 221). It would be deeply troubling if this was the conclusion. Further investigation is required.

## 13.2  Resolving the paradox

A resolution to the paradox, it will be shown, lies in adopting a Wittgensteinian–Bohrian model of educational philosophy, which will be outlined in the remainder of this chapter. By concatenating the ideas of these two philosophers – ideas which have been outlined throughout this book – this author will put forward a new model of educational philosophy which resolves this learning conundrum, proving that Meno's Paradox has profound consequences for the fruitfulness of learning theory, but not for the concept of learning itself. This age-old problem, which has plagued educational discourse since the time of Socrates and Plato, is resolvable only when one departs from the various views within educational theory which attempt to analyse, observe and measure the attributes of the learner in a manner which is atomistic, deterministic, realist and local.

These points have been variously described elsewhere in this book as fundamentally classical, Newtonian and Cartesian. That is, if the teacher–learner system is considered to be separable, then Meno's Paradox will follow. If the learner's attributes and abilities are considered to be measurement-independent, viewed rather as some pre-existing inner state, then Meno's Paradox will follow. If the learner is considered to be atomistic, entirely separate from his teacher and the community from which his teacher comes, and thus unconstrained by his training, then Meno's Paradox will follow.

In the absence of traditions, practices, forms of life, as well as the master whose responsibility it is to inculcate the novices under his[1] charge into such

---

[1]   In this chapter, the author will use the 'male' pronouns, 'he', 'him' and 'his' to refer to the collective pronouns 'he/she', 'him/her' and 'his/hers', dictated by philosophical writing style and convention.

communal conducts, Meno's Paradox is unavoidable. What this ultimately means is, precisely, that the learner's educational freedom, liberation and emancipation come only *after* their initial submission to their master's manner of acting, their efforts to assimilate his ways, their subsequent inculcation into the community whose practices and traditions they have been shown by this master, and their invitation and initiation into the *geistige* Welt, as Oakeshott called it, of human achievements and intellectual wealth.

## 13.3 The Wittgensteinian–Bohrian model of education: What it looks like, and What it means for education

Several arguments have been put forward over the course of this book that, when taken together, serve as a new vision for educational philosophy, and simultaneously offer an alternative to the constant educational theorizing which is offered at a whim within educational discourse. Indeed, what this author will draw upon in the remainder of this chapter is not an alternative theory, rather a new educational *model* which aims to dispense with the need for theory within education once and for all.

The rationale for pursuing such an approach in education in general is simple: a theory should, by its very definition, *explain* the phenomena at hand with precision, clarity and elucidation. A model of education, however, seeks not to *explain*, but merely to *describe*, offering only an attempt to demarcate between sense and nonsense within the bounds of education. As Wittgenstein opines in *On Certainty*, 'At some point one has to pass from explanation to mere description' (§189). What is particularly interesting in such a claim is that Wittgenstein is entirely clear about the lesser quality of description compared to explanations, made clear by the word 'mere'. However, such inadequacies of descriptions do not force Wittgenstein to weaken his claims that explanations are, at times, beyond the scope of the project due to the nature of what is being investigated.

This author extends this viewpoint to education, to argue that due to the constitutive uncertainty (cf. Chapter 6) which plagues the phenomena of interest to education – such as learning, thinking and so on – one must accept that explanations of such phenomena are beyond the scope of meaningful investigation, and so one must settle for 'mere description'. Consequently, educational theory is dead, and an educational model is all that is attainable.

On accepting this premise of descriptive models over explanatory theories, there is a requirement to establish some boundaries within which educational investigations can meaningfully take place, in an attempt to avoid educational paradoxes and quandaries, the most significant of which has been outlined above, namely Meno's Paradox. Therefore, the overarching philosophical rationale for establishing an educational model is akin to why the mathematician is driven to establishing the Axioms – or self-evident truths – of his modes of inquiry; that is, precisely, to make clear from the beginning what is meaningful and what is not, so as to avoid any unintelligible questions and investigations later in his studies and inquiries.

### 13.3.1 Wittgensteinian aspects of the model

There are nine core principles which have been outlined over the course of this book which make up the contributions to this author's educational model from Wittgenstein's philosophy of mind, psychology and language. They are:

1. *First-person/ third-person asymmetry*
2. *Philosophical Investigations, §304 (Chapters 4 and 11)*
3. *Philosophical Investigations, Part II, PPF §315 (Chapter 6)*
4. *Knowledge/ certainty categorial distinction*
5. *Private language argument*
6. *Rule-following paradox*
7. *Infused inner/outer relation*
8. *Language-games*
9. *Forms of life*

### 13.3.2 Bohrian/quantum physics aspects of the model

Similarly, there are a further eight core principles, primarily developed in Part 4, which come directly from Bohr's work in particular, and from within the philosophy of quantum physics in general. They are:

1. *Complementarity*
2. *Unambiguous communication*
3. *Subject/object holism*
4. *Entanglement*
5. *Non-locality*

6. *Non-separability*
7. *Irreducible uncertainty*
8. *Relationalism*

### 13.3.3 The Model: Paradox-free educational philosophy

Recall, Wittgenstein notes, 'What is you aim in philosophy? To show the fly the way out of the fly bottle' (*PI*, §309). Despite opposition from some of Wittgenstein's detractors, most notably from Karl Popper, there is no doubt some credibility in this view of philosophy. Wittgenstein's philosophy was and remains, therapeutic. It aims to rid the world of nonsense and attempts to draw definitive demarcations between intelligible and unintelligible claims. It is for this approach that this author invokes Wittgenstein's philosophical 'fly bottle' approach, to stop educationalists from – to extend the analogy – 'buzzing around' in philosophical confusion.

So, the aim here is to define a model of education, focusing particularly on the respective philosophical works of Wittgenstein and Bohr, which rids education of Meno's Paradox, and offers a meaningful description of educational interactions and exchanges, most particularly of teaching and learning, which can circumvent the inherent inadequacies of modern-day intrinsic educational theories such as neuroeducation and brain-based learning. To this end, the model which is being outlined here will subscribe to the following nine fundamental principles:

1. *An ideologically traditional educational paradigm;*
The traditional educational paradigm is historically one which holds that knowledge is central to education, and that it is the teacher who is the agent of the knowledge, whose responsibility it is to deliver the knowledge to the pupils under his charge. Rather unfortunately, traditional education is often sold – in the main by the opponents of the idea, so-called progressivists – as passive, didactic education, in which the teacher preaches to the pupil in a dogmatic, coercive and restrictive manner. In this author's own published work, one comes to appreciate that the supporters of such a view, such as Rousseau, Rogers, Dewey and Freire, are in fact mistaken when they level such a criticism at traditional education (Kitchen 2014: 40–1; 87–8).

In this model of education, however, which is being put forward in the bounds of this book, as well as in Kitchen (2014), the author is contending that traditional education need not be viewed as restrictive and coercive, rather the opposite: acting as a guide to educational emancipation and liberation.

Both the philosophies of Wittgenstein and Bohr respectively can be invoked to adopt a view of human knowledge (in the case of both Wittgenstein and Bohr) and human authority (more so in the case of Wittgenstein than of Bohr) which are steeped in traditionalism. Indeed, Wittgenstein accounts for a description of the development of human knowledge which is predicated on certainties, one such example of which, this author has argued, is one's faith in the learnings obtained from one's teacher (Kitchen 2014: 158–72). And, if education is to be about the fullest development of knowledge – which is surely a fundamental component of *any* education, regardless of one's ideological stance – then this author concludes that a traditional educational paradigm most accurately captures Wittgenstein's views on bedrock certainties forming the basis for one's ability to come to 'know' anything.

Similarly, Bohr's relational model of the description of nature accounts for an active interaction between the teacher, the pupil and the practice, which serves as the measuring device in this analogically extended model of education. Traditionalism is supported by this underpinning philosophy because the traditionalist, unlike the progressivist for example, argues in favour of the centrality of the teacher–pupil relation, as well as the importance of the practices, institutions and traditions of education, all of which are accounted for in Bohr's relational model of description, extended to education. The progressivist, however, places emphasis on the individual, positing a world in which institutions, practices and traditions are seen only as restrictive forces, which serve only to stifle creativity and drown out the voice of the student.

This view of traditionalism, however, is not one where creativity and individuality – often seen as the hallmarks of progressivist education – are scorned in practice. Rather, it is the idea that *in the first instance*, the learner is constrained in his behaviour, with a view on his eventual inculcation into the tradition – from which his teachers come – and thus his educational emancipation. Faith comes first, then knowledge, and finally the freedom to make informed choices for oneself. In this order of things, there is a dynamic infusion of traditional educational values, together with liberation and freedom, concepts which are often sold as the private possessions of the progressive ideology.

Taking Wittgenstein and Bohr together, and applying their philosophies to education, this author concludes that one is on safe ground in fitting the child to the school, and inside the educational fiduciary framework, rather than following the progressivist agenda of fitting the institutions of education to the individual child.

2. *A 'master–apprentice' style approach to teacher–pupil interactions, governed by the analogical extension of educational non-separability;*

This principle of the model follows on as a corollary of the previous discussion on adopting a traditional educational paradigm. The master–apprentice approach to teacher–pupil interactions is one which, tacitly or otherwise, assumes that the teacher is a 'master' of his subject or discipline, and as such serves as a direct representative of his practice and his community. The pupil, on the other hand, as outlined by Oakeshott (1989), for example is the natural counterpart to the teacher, a special type of learner who seeks to assimilate his master's manner of acting, in order to complete his initiation into a form of life which will allow him to inherit his intellectual wealth and freedom.

This structure of teacher–pupil interactions is clearly supported philosophically by the principle of non-separability, in particular the nature of the predicates which are ascribed to each of the players in this model; namely, the master-teacher and the apprentice-pupil. Indeed, in viewing the teacher–pupil system as governed by non-separability, one is contending that once the teacher and the pupil, initially as two separately analysable systems, have interacted with one another, the properties of each constituent part of the now interactive, entangled system are supervened upon by the properties of the system. That is, therefore, that the pupil is no longer able to act in a manner which is free from constraint from his teacher's teaching, and thus all of the pupil's future behaviours are governed by an inseparable bond to his teacher. Moreover, if there is a change in the teacher and his behaviour or his message – for example, suppose there is an alteration in the theorems of mathematics, which brings about a change in the practice of mathematics, then the teacher would clearly adopt this in his future teaching, in a manner which is different to how he may previously have done – then this will automatically bring about a simultaneous change in his pupils.

In this context, the master–apprentice structure for the teacher–pupil system and interactions is well considered, since it is not possible to consider the teacher and the pupil as existing in separately analysable states, post interaction with one another. Therefore, in order to attain access to the information and the judgement which the teacher is able to provide, the learner must assume the role of apprentice learner, ply his trade and become a part of the masterful practice from which his own master comes.

3. *A teacher-led, instructive, but participatory (i.e. not passive) pedagogy;*

Further to the non-separability governed teacher–pupil interactions, is the natural consequence of teacher-led, instructive pedagogy. This is the natural counter position to so-called 'child-centred' or 'learner-centred' pedagogy, which seeks to place the learner at the centre of educational investigation and inquiry.

However, it is this author's contention that there are two core issues with such a position, which ought to be avoided: (1) such a view of learning as atomistic and detached from any form of external constraint from, say, a teacher, results in Meno's Paradox outlined above (Kitchen 2014: 62–8) and (2) the notion that to be truly child-centred is to allow the child to act in caprice is misguided and incoherent (Kitchen 2014: 42–7; 71–3; 150–2). This author takes the alternative view that caprice leads to chaos, which in turn leads to ignorance. On the other hand, a teacher-led, instructive pedagogy gives rise to structure, which leads to constraint, giving meaning to one's actions, which results in disciplined conduct and habitual practice, consequently leading to informed modes of acting and insight.

It is important also to highlight that such a model of instructive education led by the teacher as a master in his discipline, need not necessarily be open to the criticism of being passive or submissive in the negative sense. The teacher–pupil interactions outlined in part (2) must be, by their very definition, participatory on the part of the pupil, who chooses to follow his master's manner of acting in order to gain insight and knowledge which is likely to be of use to him. Passive education, on the other hand, is a top-down model where there exists no necessity for active participation in the bottom-up direction. Within this model of teacher-led instruction, this author is contesting a dialogue which exists between teacher and pupil, with the teacher acting as the arbiter of correctness, since he is the representative of his community, practice and discipline. Such pedagogy reinstates the teacher as an expert and the school as an institution.

4. *A knowledge-based curriculum structure, with knowledge being viewed as a dichotomy of information and judgement, underpinned by trust, non-doubting conduct and instinctive certainty;*

The knowledge-based curriculum is a concept which the author argues extensively in favour of in Kitchen (2014: 47–56; 68–71; 143–6). In these passages, the author makes the argument that knowledge is a fundamental tenet of *any* education, regardless of one's philosophical or ideological dispositions. Furthermore, in keeping with Oakeshott, this author puts forward the notion that knowledge be viewed as a dichotomy of information and judgement (Kitchen 2014: 124–32), with information serving as the mere static, inert facts of knowledge, and judgement being defined as the active, participatory aspect of the knowledge edifice which is developed in the close imitation of one's master.

Such a view of education in general and learning in particular is closely aligned with and underpinned by Wittgenstein's philosophy, outlined throughout this book, in particular in relation to the ideas of the knowledge/certainty categorial distinction and the notion of trust and non-doubting behaviour serving as the foundation of one's learning endeavours.

Indeed, what seems to be true if one accepts Wittgenstein's ideas on how knowledge is formed is that knowledge presupposes the intelligibility of doubting, and that doubting itself is based on non-doubting behaviour. In other words, as Wittgenstein opines in *OC*, §415, it seems that there are bedrock certainties which are fundamental to 'all questions and all thinking'. Perhaps we might also add 'all learning' and 'all knowledge development' to this list.

What this seems to say is that there is scope for adopting a knowledge-based curriculum, in which the development of knowledge is itself based on an initial submission to the teacher's authority as a master in their subject, and to adhere to Wittgenstein's warning that 'doubting and non-doubting behaviour. There is the first only if there is the second' (*OC*, §354). Therefore, knowledge is constructed inside a fiduciary framework, as Michael Polanyi calls it, where the pupil initially trusts his teacher, with the view to inheriting his knowledge wealth. However, the parameters of such a fiduciary framework inside which knowledge can be developed demand that the pupil be mindful of the fact that he can only develop his doubting faculties once he has initially surrendered them *in the first instance*, as he places his trust in his teacher.

5. *An institution-centred, practice-driven and communitarian-focused framework, governed by the analogical extension of educational non-locality and mindful of Wittgenstein's rejection of private language and private rule-following;*

Defining education's axiological foundations as being institution-centred is predicated on the view that there is a requirement for an accepted, agreed manner of acting in one's educational enterprises. Moreover, to be practice-driven and community-focused is to give centrality to the sociological and anthropological aspects of education in offering a coherent description of meaningful educational interactions.

Perhaps more importantly, however, is the philosophical underpinning of such a conceptual framework for education, which is to be found in some of the core principles of this book, both from Wittgenstein's philosophy of mind and language and from the analogical extensions of the principle of non-locality for educational purposes.

First, consider the arguments which this author has previously outlined with reference to Wittgenstein's *private language argument* and the *rule-following paradox* (cf. Chapter 8). What was clear from these arguments was that it is unintelligible to view any meaningful action as private; that is not to say that one cannot keep one's thoughts or ruminations private, in the sense of keeping one's thoughts and ruminations *concealed*, rather that meaningful action cannot be defined (i.e. given its meaning) privately, in some private, inner realm which is epistemically inaccessible, by definition, to others. These arguments are interwoven with Wittgenstein's discussions about so-called *language-games*, by which he means any kind of meaningful linguistic or grammatical exchange between like-minded individuals. This author extends these arguments to education, to outline that the phenomena of interest to education are examples of such language-games.

For example, in relation to individual subject areas, the norms and maxims of mathematics and literature are public possessions, not private ones. One can 'do' mathematics, only when one acts in agreement with the norms and maxims of the practice of mathematics, that is, the grammar of the language-game which is 'mathematics'. One does not do mathematics privately; one is constrained in one's mathematical endeavours by the established traditions and institutions of the existing practice. Furthermore, this does not preclude that one can make original contributions to the existing practice; rather, as this author has argued previously, one is born onto a stage on which the play is already in motion, and a radical departure from this pre-existing practice is, generally speaking, uncommon. More importantly, however, is the realization that one's active engagement with the pre-existing practices and institutions of education highlights the communitarian focus of education in meaningful conduct and interactions.

What is ingenious about Wittgenstein's discussion of these concepts, however, is captured perfectly by Moyal-Sharrock in Racine and Slaney (2013: 159), where she claims:

> If grammatical rules are the product of agreement (*RFM*, §353; *Z*, §§428–30), it is not a concerted or deliberate agreement, but what Wittgenstein calls a 'peaceful agreement' (*RFM*, §323), an 'agreement in form of life' (*PI*, §241): essentially a blind agreement in our shared natural behaviour and human practices.

Moreover, Wittgenstein's philosophy is amenable to offering coherent descriptions of concepts of interest to education, most particularly of learning, which Baker and Hacker (2014: 149) capture in the following excerpt, speaking more generally about the concept of epistemic privacy and community views:

> Indeed, it has been suggested that Wittgenstein argued that it is only the possibility of correction by others and the received patterns of behaviour by others who follow the rule in a common practice that ensure the possibility of applying the distinction between thinking one is following a rule [i.e. privately] and actually following it [i.e. publically]. That is held to warrant Wittgenstein's subsequent remark that 'it is not possible to follow a rule "privately"' (*PI* §202). On this account, the concept of a (social) practice seems pivotal for Wittgenstein's reflections on meaning and the mind.

Baker and Hacker then continue,

> According to this interpretation, the fundamental issue is to clarify how *a practice yields objective standards of correctness*. ... Wittgenstein laid stress on *the importance of training, instruction and exercises in inculcating techniques* (e.g. *PI* §§189, 208). This is part of the background of our language-games (*PI*, §§179, 190). His focus is on procedures for creating or perpetuating *shared* forms of behaviour. (2014: 149–50, original emphasis, this author's underlining)

So, to build on this idea, this author is contesting that education is an arena which is committed to cultivating shared forms of behaviour. Moreover, there is a requirement for there to be what Baker and Hacker (2014) call 'objective standards of correctness', the development of which are established in training, and habitual practice. This model of education is communitarian by its very definition, and social in nature. It accounts for the past, and is therefore also anthropological. By viewing learning as a process which is the inculcation into the maxims and ideals of pre-existing forms of life, one comes to realize that one cannot possibly achieve such inculcation in the absence of the practices and institutions of education.

One final note is the corollary of these discussions in relation to the principle of educational non-locality, established in Chapter 12. Indeed, this links nicely into Wittgenstein's contributions to this discussion, given that this author has previously outlined the significance of the principle of educational non-locality with reference to how traditions and forms of life impact on the constraint placed on the learner from his previous interactions which reside in the past. In fact, the concept of non-locality, when extended analogically to education, is essentially equivalent to Wittgenstein's castigation of *private language* and *private rule-following*. Indeed, in examining Einstein's rejection of non-locality in particular, one comes to realize that it led him to adopt a local hidden variables theory of physics, which, philosophically speaking is logically equivalent to a *private language* for descriptions of phenomena of interest to physics. Since Wittgenstein rejects the notion of private language, and goes to great lengths to justify this claim (the entirety of *PI*, §§243–427, in fact), it follows by extension that the very same arguments can be used against the notion of local hidden variables theory in physics, conceptually speaking at least. Non-locality, therefore, is taken as a governing principle of the description of educational phenomena in general, and learning in particular.

6. *A relational assessment paradigm for the ascription of educational attributes and abilities;*

Relational approaches to education have been discussed in Chapters 11 and 12, so a simple reminder will suffice in this instance, to include the concept of relationalism in the author's model for education.

In essence, relationalism applies to two core aspects of educational discourse; first, it plays an important role in how one can meaningfully ascribe and describe educational attributes and abilities of the learner. Secondly, it plays a significant role in how one thinks about educational measurements, observations and assessments in general.

By accepting that relationalism governs how one measures educational attributes and abilities, the natural corollary is that one subscribes to the notion that it is unintelligible to speak meaningfully about such attributes and abilities without making reference to the measuring device in question. As a consequence, it follows that there are a great many questions which are now precluded in educational discourse, in particular those which seek to atomize the learner, and make ascriptions of attributes to him without taking into account the nature of the measurement used.

So, for example, of interest to this book, as it has been shown, theories such as neuroeducation and brain-based learning are on weak conceptual foundations, given that their measurement and observation methods fail to offer an intelligible account of the phenomena in question, such as learning, thinking and understanding, as well as abilities such as reading ability and mathematical ability. Indeed, such phenomena, when taken to be relational in nature, cannot be captured entirely (if at all) on a neural scan – the staple of the neuroeducational and brain-based learning technique – since such a form of measurement is predicated on an intrinsic, local model of measurement, with the (tacit?) assumption being that such attributes and abilities can be found within the confines of the mind or brain, that is, localized *inside* the learner, stored as *intrinsic* attributes of the individual. Therefore, since the measurement and observational methods of such theories cannot capture the relational nature of the phenomena in question, it follows that such models are fundamentally flawed.

This author, therefore, contends on the contrary, that one such model of education which captures this relational nature of the attributes and abilities is one where the practice serves as the measuring instrument, and the teacher is the agent of the community, who are the collection of like-minded masters in their subjects who, as it has been shown above in points (1)–(5), endorse, develop and disseminate the message which the teacher conveys to his pupils. In such a model of education, one never strays away from the fact that the ascription of educational attributes and abilities will *always* be made in relation to this practice.

7. *A departure from educational determinism and educational realism focusing more on the notion that teaching is a process of constraining the learner in his educational endeavours;*

It is interesting to note that Moyal-Sharrock states in Racine and Slaney (2014: 155) that according to Wittgenstein, 'We did not get to the certainty: "Humans think" from having observed humans, chairs and tables and concluded from these observations that human beings can, while chairs and tables cannot, think'. This teaches something very important in relation to educational discourse: namely, that these types of educational phenomena – thinking, learning and understanding, for example – are not empirically supported notions, rather they are certainties of nature, of a world-picture; things which are beyond the logical possibility of doubt.

Now, to be clear, this is not tantamount to Wittgenstein saying that such phenomena are beyond investigation, empirical or otherwise; rather that we do not arrive at their intelligibility or 'groundedness' by investigation, and regard their ascription to non-thinking and non-learning entities as absurd because of some kind of empirical evidence. In fact, what Wittgenstein says himself on this matter, is: 'No, experience is not the ground for our game of judging. Nor is its outstanding success' (*OC*, §131).

This realization links neatly into discussions about a castigation of determinism and realism in the analogical extensions from the conceptual foundations of quantum physics to education. Indeed, it has been argued earlier that the deterministic, realist point of view of education should be rejected given that the phenomena of interest to education are not amenable to ascriptions and descriptions which are governed by these principles.

So too, Wittgenstein seems to be saying that the questions posed by the educational determinist and the educational realist are themselves open to philosophical scrutiny. Indeed, the determinist educationalist will pose questions like 'What causes learning?' and 'Where does learning begin?' or 'If Tom does X will he be more likely to be good at Y?' However, in keeping with Wittgenstein, and with Bohr as an anti-realist, anti-determinist philosopher of physics, such questions are simply regarded as incoherent. For Wittgenstein in particular, statements of the kind 'Humans think' or 'Tom learns' are not statements of empirical fact, but instead are statements of *certainty*, which for Wittgenstein are not '*inferred* from experience' (Moyal-Sharrock in Racine and Slaney 2014: 155). There are many cases where one can see evidence of humans thinking and Tom learning, for example, but this does not mean that such phenomena are determined by some prior event nor that their meaningful definition is found in a determinist view. The 'cause and effect' model, rooted in determinism, for Wittgenstein and for Bohr, is simply unsuitable for the types of phenomena under investigation.

Likewise, a rejection of realist education is found in a castigation of the notion that the act of measurement plays no role in the meaningful ascription and description of educational attributes and abilities, as noted also in point (6). Indeed, the realist questions, akin to the determinist questions of education, are equally unappealing: 'Does Tom's ability to read exist *before* he reads (and indeed after he has finished reading)?' or 'If the human thinks, and his ability to think is something which pre-exists his demonstration of thinking (which is a supposition of educational realism), *where* does such thinking ability reside?' These questions are, as it has been shown previously, problematic. Wittgenstein and Bohr, however, in their respective rejections of realism[2] within their own disciplines, can be unified to create a conceptual model for education which does not yield such problematic questions, rather it weeds them out at source.

---

[2]  It is important to keep in mind here that Wittgenstein never quantified *himself* as an anti-realist; rather, his writings seem to fit most neatly into a rejection of realist principles. There are authors, however, (cf. Sabina Lovibond) who argue that some of Wittgenstein's work, in particular on the philosophy of religion, is in fact, *realist* in nature. The author takes the view that Wittgenstein's work is, generally speaking, anti-theoretical. Therefore, it is not that by rejecting realism that Wittgenstein is an anti-realist by default. Rather, his writings are generally anti-theoretical, but are also amenable to much of what the anti-realist would posit. This, it should be noted, is close to Bohr's position in physics also.

8. *A deflationary underpinning philosophy, focusing on descriptions rather than*
   *explanations of educational phenomena, governed by anti-Cartesian principles;*

A deflationary underpinning philosophy for education finds its justification in
Wittgenstein's various claims regarding the nature of philosophical inquiry and
description. The rationale is found in particular both in the preface and in some
of the final remarks made by Wittgenstein in his own PhD thesis, when he notes:

> What can be said at all can be said clearly, and what we cannot talk about we
> must pass over in silence. (*TLP*, preface, p. 3)
>
> When the answer cannot be put into words, neither can the question be put
> into words.
>
> The *riddle* does not exist.
>
> If a question can be framed at all, it is also *possible* to answer it. (*TLP*, §6.5)
>
> What we cannot speak about we must pass over in silence. (*TLP*, §7)

This author, therefore, adopts such a philosophical method for education also.
That is, there are many areas of education which are accessible to intelligible
questions, with worthwhile and sensible pursuits as their goal; questions which
have definitive answers, and areas of education which are open to coherent
descriptions and meaningful discussions. However, there are also many cases
when the questions which are posed within educational discourse transgress
on the bounds of sense, and as such give rise to conceptual confusion and
incoherent theories.

This author takes the view that the social sciences in general and education in
particular are open more to models of description than to theories of explanation.
Therefore, with a deflationary underpinning philosophy closely aligned to the
principles and concepts outlined in this final section of the book, one has, at
the very least, a basis for a model of education which is free from conceptual
confusion, and one which moves closer to elucidation and clarity of some of the
key aspects of education.

The role of the philosopher in education, therefore, is clearly re-established as
the one responsible for demarcating between what can be said, and said clearly,
and that which must be passed over in silence.

9. *A view of educational processes as the inculcation of learners into a form of*
   *life.*

The final point of note for this model of education is the overarching view that
education in general is viewed as a process of *inculcation* into forms of life, a
phrase invoked a great many times in Wittgenstein's philosophy to capture the

feeling of the pre-existing nature of the practices, traditions and institutions of education.

By viewing education as *inculcation* rather than, as some progressivists might argue, *indoctrination*, it is possible to view the educational processes as a form of initiation into a wealth of knowledge reserves and a bank of masterful insights into how to make a meaningful contribution to the world.

The alternative view – supported by the work of philosophers such as Rousseau, Rodgers, Freire and Dewey, for example, as well as by theorists such as Piaget and Vygotsky – that education needs to be seen as an endorsement of capricious behaviour in the pursuit of some misguided understanding of creativity and freedom is dispensed with in favour of a structured engagement between the informed adult community – who the teacher represents in an educational setting – and the as-yet-uninformed younger generation. Despite contestations from authors like Rogers, who regard such a view of education as a 'mistrust' of the learner (Kitchen 2014: 50; Rogers 1983: 186), this author argues that the learner must be stripped of some of his freedoms *in the first instance* in his educational endeavours, in order to arrive at the stage in his educational emancipation when he is free to make *informed choices* for himself, ever *constrained* by his previous engagements with his master-teacher. Such a view is predicated on one simple, yet profound belief: in order to be truly free to act, one first has to grasp what one seeks to act about.

Such vision and insight for the learner, comes only in the acknowledgement that it is the teacher who leads the way, as the guide towards and arbiter of correct conduct, which are the maxims of his role as an educational master expert, developed within and subsequently ascribed to him only once he had completed his own *inculcation* into his discipline and practice, of which he is now a master.

# Concluding Remarks

In this concluding chapter, I will outline some summarizing remarks which answer some important educational questions and debates which this book has sought to resolve.

1. *Remarks on mereology and the mereological fallacy and their relevance to education.*
The author hopes that the reader is left under no illusions that the mereological fallacy *is* directly applicable to much of the work which is being conducted within neuroeducation and brain-based learning. When claims are made within these sub-disciplines that the brain can think, learn, understand, do mathematics, read, remember, see patterns, make connections or act intelligently, it is clear that such claims are transgressions on the bounds of sense. It is not the brain which does any of these things, nor is it, as Searle (1990), for example, might suggest, the mind or the brain's mind. Rather, it is the human being.

This principle, established in detail by Bennett and Hacker (2003) for neuroscience in general, clearly applies also to neuroeducation and brain-based learning, two disciplines which are intent, at least in part, on establishing the cognitive and cogitative abilities of the brain. It has been shown that neuroeducationalists and brain-based learning theorists alike are keen to posit that the brain is the learning, thinking, intelligent organ, with efforts being made to establish that the brain is the agent of cognitive abilities that would normally be ascribed to the entire human being. The unfortunate consequence is, that since this opens these sub-disciplines to the quandaries associated with the mereological fallacy, the conceptual foundations of the work conducted within each area is no more credible than, for example, the pseudoscientific work of organology, cranioscopy or phrenology.

2. *Remarks on the first-person/third-person category error and its relevance to education.*
The unfortunate conceptual blunder made in this instance stems mainly, in this author's view, from neuroscientists and neurophilosophers making misguided

efforts to resolve the conceptual intricacies involved with the mereological fallacy, by conceding that the brain may not be the *agent* of cognitive, cogitative and psychological phenomena, but that it *is* in fact, the *locus* of such phenomena. The argument is, therefore, that although the brain may not *do* the thinking, learning or processing, for example, it is the brain *where these things take place*. So, the brain is seen as the *location* for such activities, which, as the neuroscientist might protest, is seen clearly on the neural scan.

However, such a conceptual connection is a category mistake, insomuch as what is shown on a neural scan is *brain activity*, which is physical activity, such as blood flow and oxygen levels in certain areas of the brain. Such physical activity is governed by the symmetry principle of ascription in that it is ascribed in the same manner for the first- and third-persons, respectively. Psychological phenomena and cognitive abilities, however, are governed by an *asymmetry* principle. Therefore, setting up an equivalence between the brain activity and the psychological phenomena is a category mistake.

What this tells us is precisely that whatever does take place inside the brain, cannot be logically equivalent to the psychological and cognitive phenomena which some neuroscientists and neurophilosophers suggest. This attempted sidestep of the mereological fallacy is, therefore, rejected.

3. *Remarks on the application of neuroscience to education in general.*
Due to the conceptual complications outlined in Parts (1) and (2), this author contends that neuroscience is fundamentally restricted in what it can offer, if anything of worth at all, to educational discourse.

Setting aside even the conceptual intricacies of this collaboration ever being fruitful, the reality is that from a pragmatic point of view, the adoption of neuroscience to education would seem to be unrealistic. Indeed, the nature of the neuroscientific method – the neural scan – makes the science altogether unamenable to everyday educational practice. Even the notion that teachers ought to be trained in neuroscientific principles to have a working knowledge of how the brain works – an idea endorsed heavily by the Royal Society (2011) – seems unappealing. Indeed, what value the workings of the brain would have for Mr Johnston, the mathematics teacher, when he teaches mathematics to his class, is left as a lacuna in the neuroscience–education discussion. Perhaps it might be suggested that Mr Johnston would be well-served in knowing how the brain processes mathematical concepts, or which parts of the brain carry out mathematical process. But, remember, the brain does not do these things, the human being does!

So, perhaps the neuroeducationalist might weaken his claim and suggest that it simply makes good sense to understand the brain as the organ of cognition and intelligence, just as the surgeon must know the ventricle from the atrium, and the aorta from the vena cava before he begins heart surgery. However, such a claim falls prey to the category mistake outlined in (2), since the brain is not the location of cognitive function nor intelligence, rather the location of brain activity. So, it is the remit of the neuroscientist to known the functions of the brain, not the remit of a teacher. This author thus concludes that perhaps Mr Johnston ought to be left alone to teach, which *is* his job.

4. *Remarks on the Cartesian nature of neuroscience.*
There is a great deal of philosophical writing space dedicated in Bennett and Hacker (2003), as well as in keynote lectures, Hacker (2012 and 2014), to suggest that what was once posited of the mind under Cartesian philosophy and the so-called dualist mind–body situation, is now rebranded under neuroscience as brain–body dualism. Rather than spend extended amounts of writing space extending these views with little to add, the author notes simply that what Hacker posits of neuroscience in each of these arguments is naturally extended to neuroeducation to support the thesis, which this author endorses, that such studies are steeped in Cartesian philosophy, and are, as Hacker notes, not nearly as anti-Cartesian as their supporters like to think they are.

5. *Remarks on the Newtonian nature of the psychological model adopted by neuroscience.*
Similarly, this author has extended Hacker's theses to argue that neuroscience not only is stuck in Cartesian mode, but also clings on to an outdated, and fundamentally unsuitable Newtonian model, borrowed from physics, and latterly adopted by psychology.

Furthermore, and indeed more importantly for this book, neuroeducation has subsequently fallen prey to such conceptual blunders in its foundations. In reality, this failure was deeply embedded into the foundations of psychology (cf. Lewin, 1931), and by default has found its way, tacitly or otherwise, into the foundations of neuroscience, the main manifestation being the notion that the inner realm is mental, epistemically private, hidden and linked causally to the outer, behavioural, publically accessible realm.

This view of the conceptual foundations of psychology in particular has been rejected in this book, in the main due to contributions from the respective philosophies of Wittgenstein and Bohr, whose work, when taken together, posits a more *infused* view of the inner/outer picture, with the intelligibility of meaningful ascriptions and descriptions of psychological phenomena being established in public, social and epistemically accessible criteria, from the outer world.

6. *Remarks on the inner/outer picture.*
The true nature of the inner/outer picture of psychology, and thus by this author's extension, also of education, is an *infused* relation in which neither 'realm' can be coherently spoken about or described without reference to the other. This is clearly highlighted in Wittgenstein's rejection of Cartesian dualism, as well as in the principle of first-person/third-person asymmetry, and it is also found in Bohr's discussion of subject/object holism, which he argued also applies to similar discussions within psychology. This author extends this view in this book to argue that the infused view of the inner/outer, subject/object, first-person/third-person picture is also applicable to education.

7. *Remarks on the connections between the respective philosophies of quantum physics and education.*
It has been argued in Part 4 of this book that there *is* indeed a great many overlaps between the respective philosophies of quantum physics and psychology, which this author has extended to the philosophy of education also due to the similarities which this author has established between the nature of the relata involved in each of these disciplines.

The main interest has been a unification of the ideas from Wittgenstein's philosophy of mind and Bohr's philosophy of physics, but the discussion went beyond these more restrictive parameters.

The main areas in which agreement has been established are: (1) anti-realism, (2) indeterminism, (3) holism, (4) non-locality, (5) non-separability, (6) complementarity and (7) relationalism. In each of these core postulates of quantum physics, the author has established an educational equivalent and offered insight as to how it can be invoked in a new educational model.

8. *Remarks on intrinsic and relational attributes.*
It has been shown extensively throughout that educational attributes are *not* intrinsic attributes of the learner/individual, rather they are relational attributes which are to be measured *in relation to* the existing practices, traditions and institutions of education.

This notion has been a running trend throughout the philosophical considerations of Parts 2–4 of this book, and has been developed in detail with reference to Bohr's relational attributes model of description for quantum physics.

The most notable consequence of interest to the remit of this book is that, since neuroeducation and brain-based learning theories are committed to an *intrinsic* attributes model for the attributes which they seek to measure, ascribe and describe, it follows that the conceptual foundations of these sub-disciplines are yet further called into question.

Indeed, having established that phenomena such as learning, thinking, intelligence, reading ability, mathematical ability and so on are relational in nature, it follows that when sub-disciplines such as neuroeducation and brain-based learning theory make efforts to change the manner in which these phenomena are measured, they subsequently change the *entire definition* of the phenomena in question.

To be clear, there is nothing to preclude such a shift from taking place within science in general, and there are a great many cases serving as examples of radical departures from existing scientific trends which have given rise to some of the greatest scientific breakthroughs in history. So, if neuroeducation and brain-based learning were to depart from their current intrinsic trends and adopt a relational model, perhaps might this resolve this author's concerns? However, in keeping with the mereological fallacy and the asymmetry category mistake outlined earlier, this author contests that such a scientific shift is predicated on further conceptual confusion, since the attributes and phenomena in question are properties of the entire human being *in relation to* his environment, rather than of the brain *in relation* to some interpretation of an fMRI, for example. That is, the only possible, and indeed plausible, escape for the neuroeducationalist and the brain-based learning theorist is to posit that the relation in question is not found in the interaction between the human being and the existing practices, traditions and institutions of education, rather it is the interaction of the brain with the neural scanner. So, in this view of things, learning ceases to be what is on public display and shifts to become what is shown on the neural scanner. However, remember, because of mereology, it is not the brain which learns, and further, due to the asymmetry category mistake, it is not even the brain where the learning takes place. The muddle remains unresolvable by invoking neuroscientific methods.

*9. Remarks on the connections between Wittgenstein's philosophy of mind and language, and Bohr's philosophy of physics.*

There are many connections between Wittgenstein's philosophy of mind and Bohr's philosophy of physics. Consider the following three most prominent connections:

First, there is a striking similarity between what Wittgenstein categorizes as first-person/third-person asymmetry, and what Bohr labels 'complementarity'. Indeed, as it has been shown, first-person/third-person asymmetry is also referred to as *unity of meaning with asymmetry of use*, whereas complementarity is also defined as *mutual exclusion with joint completion*. These definitions of these principles show a clear analogical connection between the respective philosophies of both philosophers.

Secondly, what this author has defined as an infused inner/outer relation in Wittgenstein's philosophy is captured in Bohr's philosophy as subject/object holism.

Also, where Wittgenstein questions the notion that abilities can be meaningfully ascribed and described in the absence of some observable action to make such ascriptions and descriptions (cf. *PI Part II*, PPF §36; *PI*, §149 (a) and (b), where Wittgenstein claims that an ability to play chess, for example, is meaningfully ascribed *only* when one plays chess and makes moves in agreement with the rules of chess), Bohr captures the very same ideas in his outline of relationalism.

## 10. *Remarks on resolving Meno's Paradox of Learning.*

The Meno's Paradox is the overarching conundrum of this book, as it drives straight to the core of the philosophical quandary facing education in general, in particular with regard to educational theory, and what is often claimed about how pupils learn.

The Meno's gives a clear and damning review of theories of learning which attempt to justify the view that learning is atomistic, and born from caprice. Moreover, it offers clear and insightful elucidation into what the problems are of viewing teaching and learning interactions as 'facilitation' on the part of the teacher, and 'discovery' on the part of the learner. Meno tells us that such a view of learning is nonsense; that is, a transgression of the bounds of sense.

The consequence of Meno *is not*, however, that learning is impossible; and this should be a great relief! Rather, it is that any theory which sustains the view that the learner can stumble around in the educational wilderness searching for answers *on their own*, in the absence of some guiding master expert, is a theory of education which is steeped in paradoxical foundations.

So, what does Meno tell us about how learning *is* possible? Quite simply, that the only way to circumvent the paradox is to abandon the constructivist view that learning is done by the individual, in a capricious, supposedly 'creative' manner, with some misunderstood notion of learner-freedom as the ultimate educational aim. Meno warns us, in no uncertain terms, that to dispense with the centrality of the teacher in the teaching–learning interaction is to dispense with the intelligibility of learning itself.

In light of Meno, the conclusion is, precisely, that one can only come to *know* anything when one puts one's faith in one's intellectual superior: namely, in one's teacher. Furthermore, it is only inside a fiduciary framework that one can access one's intellectual freedom. With a wealth of intellectual riches so readily accessible, and with an insight and a judgement which is likely to stand the test of time, it seems foolish to dispense with the teacher as a Sage. Indeed, since in this view of things the teacher holds the key to educational and intellectual emancipation, this is surely a worthwhile motivation to trust, and indeed assimilate, his manner of acting.

# Bibliography

Anscombe, G. E. M. (1985). Wittgenstein on rules and private language. *Ethics*, 95, 342–52.

Anscombe, G. E. M. (2000). *Intention*. Cambridge, MA: Harvard University Press.

Avramides, A. (2001). *Other Minds*. London: Routledge.

Baker, G. and Hacker, P. M. S. (1985). *Wittgenstein: Rules, Grammar and Necessity*. Oxford: Blackwell Publishers Ltd.

Baker, G. and Hacker, P. M. S. (2014). *Wittgenstein: Rules, Grammar and Necessity*. Revised by P. M. S. Hacker. Oxford: Blackwell Publishers Ltd.

Bax, C. (2011). *Subjectivity after Wittgenstein: The Post-Cartesian Subject and the 'Death of Man'*. London: Bloomsbury.

Bell, J. S. (1964). On the Einstein Podolsky Rosen Paradox. *Physics*, 1, 195–200.

Bell, J. S. (1981). Bertlmann's Socks and the Nature of Reality. *Le Journal de Physique Colloques*, 42(C2), 41–61.

Bell, J. S. (2010). *Speakable and Unspeakable in Quantum Mechanics* (Second Edition). Cambridge: Cambridge University Press.

Bennett, M. R. and Hacker, P. M. S. (2003). *Philosophical Foundations of Neuroscience*. Oxford: Blackwell Publishing.

Bennett, M. R., Dennett, D., Hacker, P. M. S. and Searle, J. (2007). *Neuroscience & Philosophy: Brain, Mind, & Language*. New York: Columbia University Press.

Bereiter, C. (1985). Towards a solution of the learning paradox. *Review of Educational Research*, 55(2), 210–26.

Bereiter, C. (1991). The learning paradox: Commentary. *Human Development*, 34, 294–98.

Berkeley, G. (1988). *Principles of Human Knowledge*. London: Penguin.

Bitbol, M. (2013). Bohr's Complementarity and Kant's Epistemology. *Séminare Poincaré*, XVII, 145–66.

Bohr, N. (1934/1987). *The Philosophical Writings of Niels Bohr: Volume 1 – Atomic Theory and the Description of Nature*. Woodbridge: Ox Bow Press.

Bohr, N. (1958/1987). *The Philosophical Writings of Niels Bohr: Volume 2 – Essays 1933 – 1957 on Atomic Physics and Human Knowledge*. Woodbridge: Ox Bow Press.

Bohr, N. (1963/1987). *The Philosophical Writings of Niels Bohr: Volume 3 – Essays 1958 – 1962 on Atomic Physics and Human Knowledge*. Woodbridge: Ox Bow Press.

Bohr, N. (1963/1987). *The Philosophical Writings of Niels Bohr: Volume 4 – Essays 1958 – 1962 on Causality and Complementarity*. Woodbridge: Ox Bow Press.

Bohr, N. (1981–2008). *Niels Bohr Collected Works: Volumes 1-13*. L. Edited by L. Rosenfeld, J. Rud Nielsen, Ulrich Hoyer, Erik Rüdinger, Klaus Stolzenburg, Jørgen Kalckar, Finn Aaserud, Jens Thorsen, Sir Rudolf Peierls, David Favrholdt. Amsterdam: North-Holland Publishing Company.

Bohr, N. (2010). *Atomic Physics and Human Knowledge*. New York: Dover.

Bohr, N. (2011). *Atomic Theory and the Description of Nature*. Cambridge. Cambridge University Press.

Bruer, J. (1994). Classroom Problems, School Culture, and Cognitive Research. In K. McGilly (ed.), *Classroom Lessons: Integrating Cognitive Theory and Classroom Practice*. Cambridge, MA: The MIT Press.

Bruer, J. (November 1997). Education and the brain: a bridge too far. *Educational Researcher*, 26(8), 4–16.

Bruer, J. (May 1998a). Let's put brain science on the back burner. *NASPP Bulletin*, 82(598), 9–20.

Bruer, J. (November 1998b). Brain science, brain fiction. *Educational Leadership*, 56(3), 14–19.

Bruer, J. (1999b). *The Myth of the First Three Years: A New Understanding of the Early Brain Development and Lifelong Learning*. Detroit: Free Press.

Bruer, J. (November 2002b). Avoiding the pediatrician's error: how neuroscientists can help educators (and themselves). *Nature Neuroscience*, 5(11), 1031–3.

Caine, R. N. and Caine, G. (1991). *Making Connections: Teaching and the Human Brain*. Alexandria, VA: Association for Supervision and Curriculum Development.

CCEA (2003a). *Is the Curriculum Working? Summary of the Key Stage 3 Phase of the Northern Ireland Curriculum Cohort Study*. Belfast: Council for the Curriculum, Examinations and Assessment.

CCEA (2003b). *Proposals for Curriculum and Assessment at Key Stage 3. Part 1: Background Rationale and Detail*. Belfast: Council for the Curriculum, Examinations and Assessment.

CCEA (2003c). *Proposals for Curriculum and Assessment at Key Stage 3. Part 2: Discussion Papers and Case Studies*. Belfast: Council for the Curriculum, Examinations and Assessment.

CCEA (2003d). *Proposals for Curriculum and Assessment at Key Stage 3. Teacher's Folder*. Belfast: Council for the Curriculum, Examinations and Assessment.

CCEA (2006). Response to Dr Hugh Morrison's criticisms of the revised curriculum and assessment proposal including the pupil profile. Available online at www.ccea.org.uk

CCEA (2007a). *The Statutory Curriculum at Key Stage 3: Rationale and detail*. Belfast: Council for the Curriculum, Examinations and Assessment.

CCEA (2007b). *Active Learning and Teaching Methods for Key Stage 3*. Belfast: Council for the Curriculum, Examinations and Assessment.

Churchland, P. S. and Sejnowski, T. J. (2000). The Computational Brain: Anatomical and Physiological Techniques, in Robert Cummins and Denise Dellarosa Cummins (eds), *Minds, Brains, and Computers: An Anthology*. Oxford: Blackwell Publishers.

Crane, T. (1995/2003). *The Mechanical Mind*. London: Routledge.

de Jong, T., van Gog, T., Jenks, K., Manlove, S., van Hell, J. G., van Merriënboer, J. J. G., van Leeuwen, T. and Boschloo, A. (2008). *Explorations in Learning and the Brain: On the potential of Cognitive Neuroscience for Educational Science*. The Hague (NL): Netherlands Organisation for Scientific Research.

Dellarosa Cummins, D. (2000). A History of Thinking, in Robert Cummins and Denise Dellarosa Cummins (eds), *Minds, Brains, and Computers: An Anthology*. Oxford: Blackwell Publishers.

Dennett, D. C. (1987). *The Intentional Stance*. Cambridge, MA: MIT Press.

Dennett, D. C. (1991). *Consciousness Explained*. Boston: Little, Brown and Company.

Dennett, D. C. (1998). *Brainchildren: Essays on Designing Minds*. Cambridge, MA: The MIT Press.

Einstein, A., Podolsky, B. and Rosen, N. (1935). Can Quantum-Mechanical Description of Physical Reality Be Considered Complete? *Physical Review*, 7, 777–80.

Favrholdt, D., ed. (1999). *Niels Bohr Collected Works (Volume 10)*. Amsterdam: Elsevier Science B.V.

Faye, J. (1991). *Niels Bohr: His Heritage and Legacy: An Anti-Realist View of Quantum Mechanics*. Dordrecht: Kluwer.

Faye, J. and Folse, H. J. (1994). *Niels Bohr and Contemporary Philosophy*. Dordrecht: Kluwer Academic Publishers.

Feynman, R. (1985). *QED: The Strange Theory of Light and Matter*. Princeton: Princeton University Press.

Fine, A. (1989). Do Correlations need to be Explained? in J. T. Cushing and E. McMullin (eds), *Philosophical Consequences of Quantum Theory: Reflections on Bell's Theorem*, 175–94. Notre Dame, IN: University of Notre Dame Press.

Gazzaniga, M. S., Ivry, R. B. and Mangun, G. R. (1998). *Cognitive Neuroscience: The Biology of the Mind*. New York: W. W. Norton & Company.

Geake, J. (2005). Educational neuroscience and neuroscientific education: in search of the mutual middle-way. *Research Intelligence: News from the British Educational Research Association*, 92, 10–13.

Geake, J. (2010). Neuromythologies in education, in Paul Howard-Jones (ed.), *Education and Neuroscience: Evidence, Theory and Practical Application*, 5–15. Oxon: Routledge.

Geake, J. and Cooper, P. (2003). Cognitive Neuroscience: Implications for Education? *Westminster Studies in Education*, 26(1), 7–20.

Glock, H-J. (1996). *A Wittgenstein Dictionary*. Oxford: Blackwell Publishers.

Goswami, U. (2006). Neuroscience and Education: From Research to Practice?. *Nature Reviews Neuroscience*, 7, 406–13.

Goswami, U. (2010). Reading, Dyslexia and the Brain, in Paul Howard-Jones (ed.), *Education and Neuroscience: Evidence, Theory and Practical Application*, 16–29. Oxon: Routledge.

Gulpinar, M. A. (2005). The Principles of Brain-based Learning and Constructivist Models in Education. *Educational Sciences: Theory & Practice*, 5(2), 299–306.

Hacker, P. M. S. (1986). *Insight and Illusion: Themes in the philosophy of Wittgenstein.* Oxford: Clarendon Press.

Hacker, P. M .S. (1993). *Wittgenstein Meaning and Mind.* Oxford: Blackwell Publishers.

Hacker, P. M. S. (1997). *Wittgenstein on Human Nature.* London: Phoenix.

Hacker, P. M. S. (2010). *Human Nature: The Categorial Framework.* Oxford: Blackwell Publishers.

Hacker, P. M. S. (2012). *Are Persons Brains? The Challenge of Crypto-Cartesianism.* Keynote lecture, IRC Conference at St Anne's College, Oxford.

Hacker, P. M. S. (2014). *What can the Brain Teach us about the Mind?* Keynote lecture, in the debate at the LSE with Professor Ray Nolan and Professor Nikolas Rose.

Harré, R. and Tissaw, M. (2005). *Wittgenstein and Psychology.* Burlington, VT: Ashgate.

Heisenberg, W. (1958). *Physics and Philosophy.* New York: Harper & Row.

Held, C. (1995). *Bohr and Kantian Idealism.* Lecture at The Eighth International Kant Conference, Memphis, Marquette University. Available online at: http://www2.uni-erfurt.de/wissenschaftsphilosophie/Held/Bibliographie/bohr-and-kantien.pdf.

Honner, J. (1987). *The Description of Nature: Niels Bohr and the Philosophy of Quantum Physics.* Oxford: Oxford University Press.

Howard-Jones, P. (January 2014). Neuroscience and Education: A Review of Educational Interventions and Approaches Informed by Neuroscience. Available online at https://educationendowmentfoundation.org.uk/public/files/Publications/EEF_Lit_Review_NeuroscienceAndEducation.pdf

Howard-Jones, P., Pollard, A., Blakemore, S. J., Rogers, P., Goswami, U., Butterworth, B., et al. (2007). Neuroscience and Education: Issues and Opportunities. A commentary by the Teaching and Learning Research Programme. Available online at www.tlrorg

Howse, P. (February 2015). Where teachers' brains detect student confusion. *BBC Online.* Retrieved from http://www.bbc.co.uk/news/education-31503265

Hume, D. (1739). *A Treatise of Human Nature.* London: John Noon.

Ismael, J. and Schaffer, J. (2014). Quantum Holism as Common Ground. Available online at: http://www.jonathanschaffer.org/quantumholism.pdf.

James, W. (1890). *The Principles of Psychology (Volume 1).* New York: Dover Publications.

Jeans, J. (1930). *The Mysterious Universe.* Cambridge: The University Press

Jensen, E. (2008). *Brain-based Learning: The New Paradigm of Teaching.* Thousand Oaks, CA: Corwin Press.

Johnston, P. (2000). *Wittgenstein: Rethinking the Inner.* Oxon: Routledge.

Kalckar, J., ed. (1985). *Niels Bohr Collected Works (Volume 6).* Amsterdam: Elsevier Science B.V.

Katsumori, M. (2011). *Niels Bohr's Complementarity: Its Structure, History, and Intersections with Hermeneutics and Deconstruction.* London: Springer.

Kenny, A. (2004). *The unknown God: Agnostic Essays.* London: Continuum.

Kitchen, W. H. (2014). *Authority and the Teacher.* London: Bloomsbury.

Kochen, S. and Specker, E. (1967). The Problem of Hidden Variables in Quantum Mechanics. *Journal of Mathematics and Mechanics*, 17, 59–87.

Kohler, W. (1938). *The Place of Mind in a World of Facts*. New York: Liveright.

Kripke, S. A. (1982). *Wittgenstein on Rules and Private Language*. Oxford: Blackwell.

Kuhn, T. S. (1962). *The Structure of Scientific Revolutions*. Chicago: University of Chicago Press.

Kuhn, T. S. (1963). The Function of Dogma in Scientific Research, in A. C. Crombie (ed.), *Scientific Change*, 347–69. London: Heinemann.

Lee, J. (March 2013). Open your mind to the teachings of neuroscience. Available online at: http://www.tes.co.uk/article.aspx?storycode=6322171

Malcolm, N. (1971). *Problems of Mind: Descartes to Wittgenstein*. New York: Harper.

Malcolm, N. (1977). *Memory and Mind*. Ithaca: Cornell University Press.

Malcolm, N. (1986). *Wittgenstein: Nothing is Hidden*. Oxford: Blackwell.

Malcolm, N. (1995). *Wittgensteinian Themes; Essays 1978-1989*. Ithaca: Cornell University Press.

McDowell, J. (1998). *Mind, Value and Reality*. Cambridge, MA: Harvard University Press.

McGinn, C. (1984). *Wittgenstein on Meaning*: Oxford: Blackwell.

McGinn, C. (1996). *The Character of Mind: An Introduction to the Philosophy of Mind*. Oxford: Oxford University Press.

McGinn, M. (1997). *Wittgenstein and the Philosophical Investigations*. London: Routledge.

Mitchell, M. T. (2006). *Michael Polanyi, The Art of Knowing*. Wilmington, Delaware: T. Kenneth Cribb Jr.

Monk, R. (2005). *How to Read Wittgenstein*. London: Granta Books.

Morrison, H. (2006). Brain Based Learning? *Fortnight*, 440, 9–10.

Morrison, H. (2013). A fundamental conundrum in psychology's standard model of measurement and its consequences for PISA global rankings. Available online at: http://paceni.wordpress.com/2013/11/28/the-paper-which-topples-oecd-pisa-2012/

Morrison, H. (2014). Knewton claims of adaptive learning have no scientific merit. Available online at: https://paceni.wordpress.com/2014/04/08/knewton-claims-of-adaptive-learning-have-no-scientific-merit/

Moyal-Sharrock, D. (2007). *Understanding Wittgenstein's On Certainty*. New York: Palgrave Macmillan.

Mundasad, S. (July 2015). 'Celebrity mind game' may reveal clues to memory. Available online at: http://www.bbc.co.uk/news/health-33343511

Murdoch, D. (1987). *Niels Bohr's Philosophy of Physics*. Cambridge: Cambridge University Press.

Oakeshott, M. (1975). *On Human Conduct*. Oxford: Oxford University Press.

Oakeshott, M. (1989). *The Voice of Liberal Learning*. New Haven: Yale University Press.

Oakeshott, M. (1991). *Rationalism in Politics and other Essays*. Indiana: Liberty Fund, Inc..

Oppenheimer, R. (1955, September 4). *Analogy in Science*. Paper presented at the 63rd Annual Meeting of the American Psychological Association, San Francisco, CA.

Panjvani, C. (2008). Rule-following, Explanation-transcendence, and Private Language. *Mind*, 117, 303–28.

Pears, D. (1985). *Wittgenstein*. London: Fontana Press.

Pickering, S. J. and Howard-Jones, A. (2007). Educators' views of the Role of Neuroscience in Education: A Study of UK and International Perspectives. *Mind, Brain and Education*, 1(3).

Place, U. T. (2000). Is Consciousness a Brain Process? in Robert Cummins and Denise Dellarosa Cummins (eds), *Minds, Brains, and Computers: An Anthology*. Oxford: Blackwell Publishers.

Pockett, S., Banks, W. and Gallagher, S. (2009). *Does Consciousness Cause Behavior?* Cambidge, MA: The MIT Press.

Polanyi, M. (1946). *Science, Faith and Society*. London: Oxford University Press.

Polanyi, M. (1958). *Personal Knowledge: Towards a Post-Critical Philosophy*. Chicago: The University of Chicago Press.

Polanyi, M. (1969). *Knowing and Being*. London: Routledge and Keegan Paul.

Polanyi, M. (1983). *The Tacit Dimension*. s.l.: Doubleday & Company.

Polkinghorne, J. (1996). *Beyond Science*. Cambridge: Cambridge University Press.

Purdy, N. (2008). Neuroscience and Education: How Best to Filter out the Neurononsense from our Classrooms? *Irish Educational Studies*, 27(3), 197–208.

Purdy, N. and Morrison, H. (2009). Cognitive Neuroscience and Education: Unravelling the Confusion. *Oxford Review of Education*, 35(1), 99–109.

Purves, D., Cabeza, R., Huettel, S. A., LaBar, K. S., Platt, M. L. and Woldorff, M. G. (2013). *Principles of Cognitive Neuroscience*. Sunderland, MA: Sinauer Associates, Inc.

Putnam, H. (1960). Minds and Machines, in Sidney Hook (ed.), *Dimensions of Mind: A Symposium*. New York: New York University Press.

Putnam, H. (November 1973). Meaning and Reference. *The Journal of Philosophy*, 70(19), 699–711.

Putnam, H. (1975). *Mind, Language and Reality: Philosophical Papers. Volume 2*. Cambridge: Cambridge University Press.

Putnam, H. (1991). *Representation and Reality*. Cambridge, MA: MIT Press.

Putnam, H. (2000). Minds and Machines, in Robert Cummins and Denise Dellarosa Cummins (eds), *Minds, Brains, and Computers: An Anthology*. Oxford: Blackwell Publishers.

Racine, T. and Slaney, K. L. (2013). *A Wittgensteinian Perspective on the Use of Conceptual Analysis in Psychology*. Hampshire: Palgrave Macmillan.

Robinson, D. (2007). Still Looking: Science and Philosophy in Pursuit of Prince Reason, in M. Bennett, D. Dennett, P. Hacker and J. Searle (eds), *Neuroscience and Philosophy: Brain, Mind, and Language*, 171–93. New York: Columbia University Press.

Rowlands, M. (2003). *Externalism: Putting Mind and World Back Together again.* Guildford: McGill-Queen's University Press.

The Royal Society (2011). *Brain Waves Module 2: Neuroscience: Implications for Education and Lifelong Learning.* London: The Royal Society.

Ryle, G. (2009). *The Concept of Mind (60th Anniversary Edition).* Oxon: Routledge.

Searle, J. R. (1990). Is the Brain's Mind a Computer Program? *Scientific America,* January Edition, 26–31.

Searle, J. R. (2000). Minds, Brains, and Programs, in Robert Cummins and Denise Dellarosa Cummins (eds), *Minds, Brains, and Computers: An Anthology.* Oxford: Blackwell Publishers.

Searle, J. R. (2002). End of the Revolution. *New York Review of Books,* XLIX (3).

Sejnowski, T. J., Koch, C. and Churchland, P. S. (2000). Computational Neuroscience, in Robert Cummins and Denise Dellarosa Cummins (eds), *Minds, Brains, and Computers: An Anthology.* Oxford: Blackwell Publishers.

Shimony, A. (1997). On mentality, Quantum Mechanics and the Actualization of Potentialities, in R. Penrose (ed.), *The Large, the Small and the Human Mind,* 144–60. Cambridge: Cambridge University Press.

Stapp, H. P. (1993). *Mind, Matter, and Quantum Mechanics.* Berlin: Springr-Verlag.

Stenholm, S. (2011). *The Quest for Reality: Bohr and Wittgenstein – two Complementary Views.* Oxford: Oxford University Press.

Stern, G. (1991). Models of Memory: Wittgenstein and Cognitive Science. *Philosophical Psychology,* 4(2), 203–18.

Stern, G. (2004). *Wittgenstein's Philosophical Investigations: An Introduction.* Cambridge: Cambridge University Press

Stewart, W. (February 2014). Neuroscience is a no-brainer? Think again, expert says. Available online at: http://www.tes.co.uk/article.aspx?storycode=6408602

Stroll, A. (1994). *Moore and Wittgenstein on Certainty.* New York: Oxford University Press.

Suter, R. (1989). *Interpreting Wittgenstein: A Cloud of Philosophy, a Drop of Grammar.* Philadelphia: Temple University Press.

Suter, R. (1990). Characteristics of Criteria. Special issue of *Revista de Filosofia,* 195–202.

Swart, T., Chisholm, K. and Brown, P. (2015). *Neuroscience for Leadership.* London: Palgrave McMillan.

Sylwester, R. (1995). *A Celebration of Neurons: An Educator's Guide to the Human Brain.* Alexandria, VA: Association for Supervision and Curriculum Development.

Sylwester, R. (1997b, February). The Neurobiology of Self-esteem and Aggression. *Educational Leadership,* 54(5), 75–9.

Sylwester, R. (1998a, November). Art for Brain's Sake. *Educational Leadership,* 56(3), 36–40.

Tallis, R. (2011). *Aping Mankind.* Durham: Acumen.

ter Hark, M. R. M. (1990). *Beyond the Inner and the Outer: Wittgenstein's Philosophy of Psychology.* Dordrecht: Kluwer Academic Publishers.

Thompson, H. and Maguire, S. (2001). *Mind your Head: Get to know your Brain and how to Learn*. Antrim: North Eastern Education and Library Board (NEELB).

Tulving, E. (2007). Coding and Representation: Searching for a Home in the Brain, in H. L. Roediger III, Y. Dudai and S. M. Fitzpatrick (eds), *Science of Memory: Concepts*, 65–68. New York: Oxford University Press.

Turing, A. M. (2000). Computing Machinery and Intelligence, in Robert Cummins and Denise Dellarosa Cummins (eds), *Minds, Brains, and Computers: An Anthology*. Oxford: Blackwell Publishers.

Vygotsky, L. S. (1987). *The Collected Works of L.S. Vygotsky*. Volume 1: Problems of general psychology, including the volume *Thinking and Speech*, R. W. Rieber and A. S. Carton (eds), N. Minick (trans). New York: Plenum Press.

Vygotsky, L. S. (1997). *Educational Psychology*. Boca Ration, FL: St Lucie Press.

Ward, H. (January 2014). From Brain Scan to Lesson Plan? Millions for Neuroscience Research in UK Classrooms. Available online at: http://news.tes.co.uk/news_blog/b/weblog/archive/2014/01/07/millions-for-neuroscience-research-in-uk-classrooms.aspx.

Waterfield, R., trans. (2005). *Plato: Meno and other Dialogues*. Oxford: Oxford University Press.

Webb, J. (July 2015). Peeking into the Brain's Filing System. Available online at: http://www.bbc.co.uk/news/science-environment-33380677

Whitaker, A. (1996). *Einstein, Bohr and the Quantum Dilemma*. Cambridge: Cambridge University Press.

Wick, D. (1995). *The Infamous Boundary*. New York: Copernicus.

Williams, M. (1999). *Wittgenstein, Mind and Meaning: Towards a Social Conception of Mind*. London: Routledge.

Winch, C. (2008). *Learning how to Learn: A Critique*. Keynote lecture, Philosophy of Education Society of Great Britain Annual Conference.

Wisdom, J. (1967). A Feature of Wittgenstein's Technique, in K. T. Fann (ed.), *Ludwig Wittgenstein: The Man and His Philosophy*, 353–65. New York: Delta.

Wittgenstein, L. (1922). *Tractatus Logico-Philosophicus*. Translated by D. F. Pears and B. F. McGuinness. Oxon: Routledge & Kegan Paul.

Wittgenstein, L. (1956). *Remarks on the Foundations of Mathematics*. Cambridge, MA: MIT Press.

Wittgenstein, L. (1967). *Zettel*. Edited by G. E. M. Anscombe and G. H. von Wright. Translated by G. E. M. Anscombe. Oxford: Blackwell.

Wittgenstein, L. (1969). *On Certainty*. Edited by G. E. M. Anscombe and G. H. von Wright. Translated by Denis Paul and G. E. M. Anscombe. Oxford: Blackwell.

Wittgenstein, L. (1976). Cause and Effect: Intuitive Awareness. Edited by R. Rhees. Translated by P. Winch, *Philosophia*, 6, 392–445.

Wittgenstein, L. (1980a). *Remarks on the Philosophy of Psychology: Volume I*. Edited by G. E. M. Anscombe and G. H. von Wright. Translated by G. E. M. Anscombe. Oxford: Blackwell.

Wittgenstein, L. (1980b). *Remarks on the Philosophy of Psychology: Volume II*. Edited by G. E. M. Anscombe and G. H. von Wright. Translated by G. E. M. Anscombe. Oxford: Blackwell

Wittgenstein, L. (1980c). *Culture and Value*. Edited by G. H. von Wright and H. Nyman. Translated by P. Winch. Oxford: Blackwell.

Wittgenstein, L. (1982a). *Last Writings on the Philosophy of Psychology: Volume I*. Edited by G. H. von Wright and Heikki Nyman. Translated by C. G. Luckhardt and Maximilian A. E. Aue. Oxford: Blackwell.

Wittgenstein, L. (1982b). *Wittgenstein's Lectures: Cambridge 1930-1932*. Edited by D. Lee. Chicago: Chicago University Press.

Wittgenstein, L. (1982c). *Wittgenstein's Lectures: Cambridge 1932-1935*. Edited by A. Ambrose. Chicago: Chicago University Press.

Wittgenstein, L. (1992). *Last Writings on the Philosophy of Psychology: Volume II*. Edited by G. H. von Wright and Heikki Nyman. Translated by C. G. Luckhardt and Maximilian A. E. Aue. Oxford: Blackwell.

Wittgenstein, L. (2009). *Philosophical Investigations (4th Revised Edition)*. Edited by P. M. S. Hacker and Joachim Schulte. Translated by G. E. M. Anscombe and M. S. Hacker and Joachim Schulte. Oxford: Wiley-Blackwell.

Wright, A. (2010). *Brain Scanning Techniques (CT, MRI, fMRI, PET, SPECT, DTI, DOT)*. Retrieved from http://www.cerebra.org.uk

Wright, C. (2001). *Rails to Infinity*. Cambridge, MA: Harvard University Press.

Zadina, J. N. (2015). The Emerging Role of Educational Neuroscience in Educational Reform. *Psicología Educativa*, 21, 71–7.

# Index